D0875742

Animals under the Swastika

Animals under the Swastika

J. W. Mohnhaupt

Translated by John R. J. Eyck

THE UNIVERSITY OF WISCONSIN PRESS

Publication of this book has been made possible, in part,
through support from the Anonymous Fund of the College of Letters and
Science at the University of Wisconsin–Madison. The translation of this work
was supported by a grant from the Goethe-Institut.

The University of Wisconsin Press
728 State Street, Suite 443
Madison, Wisconsin 53706
uwpress.wisc.edu

Gray's Inn House, 127 Clerkenwell Road
London ECIR 5DB, United Kingdom
eurospanbookstore.com

Printed in the United States of America
This book may be available in a digital edition.

Library of Congress Cataloging-in-Publication Data

Names: Mohnhaupt, Jan, 1983- author. | Eyck, John R. J., translator.
Title: Animals under the swastika / J. W. Mohnhaupt ; translated by John R. J. Eyck.
Other titles: Tiere im Nationalsozialismus. English
Description: Madison, Wisconsin : The University of Wisconsin Press, [2022] |
Originally published by Carl Hanser Verlag under the title Tiere im Nationalsozialismus,
copyright ©2020 Carl Hanser Verlag GmbH & Co.
KG, München. | Includes bibliographical references and index.
Identifiers: LCCN 2021049409 | ISBN 9780299338008 (hardcover)
Subjects: LCSH: National socialism. | Human-animal
relationships—Germany—History—20th century. |
Animals—Germany—History—20th century. | Animals—Government
policy—Germany—History—20th century.
Classification: LCC DD256.5 .M56413 2022 | DDC 335.6—dc23/eng/20220120
LC record available at https://lccn.loc.gov/2021049409

To my grandfathers
HANS and HANS

CONTENTS

Animals under the Swastika

Prologue

The World behind the Wire

What an astonishing hierarchy among animals! Man sees them according to how he stole their qualities.

— Elias Canetti, *The Human Province*

I n the shadow of beech and oak trees, on the north slope of a mountain, in the middle of Germany, there was once a zoo. It was just a very small zoo, but, in addition to a koi pond, a monkey island, and bird aviaries, it also housed a bear den measuring about ten by fifteen meters. All around there were benches for the men who took their lunch breaks here. Some of them would tease the monkeys; others would watch two young brown bears that would get up on their hind legs, trying to push through the pen with their front paws. As he wrote in an official communiqué, Karl Koch had the little zoo built to provide his employees with "diversion and entertainment" and "to present the animals in all their beauty and unique character, which the workers would otherwise hardly have occasion to observe or acquaint themselves with in the wild."[1] The men who built the zoo were "behind the wire," as Koch called the three-meter-high, three-kilometer-long electric fence. Behind it stretched a wide, sloping expanse. In the summer, it was dry and dusty; in winter, icy winds swept over it. Endless rows of wooden barracks stood here, side by side.

The Buchenwald Zoological Garden, as the small animal park was officially called, and the concentration camp with the same name were only a stone's throw away from each other. From the crematorium to the bear den, there were maybe ten, at most fifteen, paces. At one time, the electric wire fence between them was the border between the Buchenwald of the prisoners and that of the guards, supervisors, and civilian workers. It

3

constituted the boundary between humans and animals on the one side and "Untermenschen" ("subhumans") on the other. The fence kept worlds apart.

Today very little recalls the zoo anymore, which the SS had built in 1938, as a "recreation area" right next to the camp. In 1993, the Buchenwald Memorial began uncovering what remained of it. A few foundation walls were still preserved, including those of the bear den, which had withstood the test of time under the brush and foliage. "We wanted to make the zoo visible again," says Rikola-Gunnar Lüttgenau, spokesperson for the memorial. Supposedly it was for didactic reasons above all else: "It is disconcerting to imagine the Nazis visiting the zoo with their children and watching the animals while people were dying right next door. Because you recognize that part of your own way of being normal, like going to a zoo, can also be part of a world where you do not feel you belong at all."[2]

Anybody visiting the ruins of the zoo today who walks around the low brick wall and the remains of the climbing rock will notice this erstwhile idyll's immediate proximity to the Buchenwald concentration camp. The zoo obviously served as a kind of smoke screen, a divider that in fact hid nothing but just shielded the supervisors' area from the prisoners' camp. "The SS prettied it up for themselves," Lüttgenau says.[3]

Until recently, research into the camp zoo has been rather scant, although it shows up frequently in historical descriptions as well as in newspaper articles and in drawings made by former prisoners.[4] It also inspired author Jens Raschke to write a children's play that he titled *Was das Nashorn sah, als es auf die andere Seite des Zauns schaute* (What the rhinoceros saw when it looked at the other side of the fence). It relates an anecdote, found in an eyewitness report, according to which a rhinoceros supposedly lived at the Buchenwald zoo, at least for a short time.[5] Sabine Stein runs the memorial's archives and knows the story, though there is no evidence for it: "Time and again, I've asked survivors about it when they came for memorial services," Stein says, "but no one could recall any rhino."[6]

While the rhino is likely a legend, the Buchenwald zoo was real and, moreover, not the only one of its kind. In the Treblinka extermination camp, too, there was a dovecote, as well as cages with foxes and other wild animals, for the diversion of the guards.[7]

The animals, which came mostly from the zoo in Leipzig, had been purchased with the meager wages that the inmates received for their forced labor in the surrounding factories, workshops, and quarries.[8] When animals were injured, it was frequently blamed on the prisoners. When one died, they also had to pay for its replacement in the form of a "voluntary assessment."[9]

Posts as zookeepers were coveted, especially those for the bear den because those employed there always had access to meat and honey. Nobody, once they had worked there, ever wanted to give up the position. Hans Bergmann, too, was willing to risk a whole lot for it. In October 1939, this Jewish prisoner wrote a letter to the camp's chief warden and "most obediently" asked him to be allowed to work with the bears again, because the current keeper, a Roma inmate, supposedly could not cope on his own with the four animals, including the pregnant female named Betty. Bergmann felt that everything had to be done to help her cubs pull through. Moreover, he noted, "I am very attached to the animals and am certainly convinced that, together with the gypsy, I can muster all four bears, plus raise the cubs in a few weeks."[10]

The guards themselves preferred to employ Sinti and Roma for work with the bears, as Lüttgenau confirms. The "gypsies"—the conventional, racist term used for the Roma at the time—hired themselves out as itinerant artists and performers and frequently put on shows with dancing bears. "Therefore the SS obviously assumed that they were 'inherently' able to get along particularly well with these animals," Lüttgenau says.[11]

The camp warden forwarded Bergmann's letter to his superior, Karl Koch, who was the commandant of the Buchenwald concentration camp. He lived on the south slope of the mountain, on the sunny side, where he additionally had the SS-Falkenhof built, a courtyard of sorts with cages for owls, eagles, and ravens, as well as enclosures for wolves, deer, and wild boar. Whereas the zoo next to the camp fence was reserved for the guards and civilian workers at Buchenwald only, the people of the nearby town of Weimar were allowed to visit the Falkenhof on weekends. They also knew about the zoo, however, because the SS marketed postcards in town depicting the brown bears of Buchenwald at play, with a caption that read "Bear Den. Buchenwald Zoo."[12]

Ilse Koch, the wife of the camp commandant, also went for strolls with their children through the small animal park. And their way always took them along the electric wire fence. Though it was otherwise strictly forbidden to take photographs there, there are images in their family album showing Karl Koch feeding and petting the animals with his son Artwin.[13] A few years later, Ilse Koch would stand before an American military tribunal and claim to have noticed neither the fence nor the camp behind it.[14]

Karl Koch was concerned that the animals not be disturbed, and he issued orders prohibiting "any feeding or teasing whatsoever."[15] Anyone who did something to the animals, however minor—who climbed over the fence onto the rocks in the bear den, say, or who even leaned against one of the cages—could count on being punished. That held true for the SS squads as well. After all, the animals were supposed to thrive, and so the prisoner Bergmann's request must have seemed compelling to Koch. He therefore endorsed his petition to be employed as a bear keeper. Next to his signature, though, he also left the following note: "*If any cub dies, punish harshly.*"[16]

Of *Herrentiere* and *Menschentiere*

It would be all too easy to discount Karl Koch's concern for the well-being of his zoo animals as an unsettling anecdote and ignore the possibility that it had implications, were it not part of a systematic shifting of boundaries, one that turned sought-after animals into *Herrentiere* (master animals) and arbitrarily reduced people to *Menschentiere* (human animals). The protection of animals and crimes against humanity did not present any contradiction for leading National Socialists. On the contrary, the Nazis even felt they belonged to a "moral elite." As Heinrich Himmler boasted in his 1943 speech in Posen (now Poznań), "Whether or not 10,000 Russian women collapse with exhaustion while digging an anti-tank ditch concerns me only insofar as the anti-tank ditch is being dug for Germany. We will never be brutal and callous unless it is necessary: that is obvious. We Germans, who alone on this earth have a decent attitude to animals, will of course adopt a decent attitude to these human animals."[17]

Rudolf Hoess, the camp commandant of Auschwitz, also emphasized the special relationship that supposedly had connected him to animals since his childhood. Horses, in particular, appealed to him.[18] During his time at Auschwitz, he sought their closeness most of all when it was no longer possible for him to justify the killing he engaged in every day, on the grounds that he was doing his duty and being obedient: "I had to go on with this process of extermination. I had to continue this mass murder and coldly to watch it, without regard for the doubts that were seething deep inside me," he wrote in the memoirs he composed after the war during his prison term in Poland. "If I was deeply affected by some incident, I found it impossible to go back to my home and my family. I would mount my horse and ride until I had chased the terrible picture away. Often, at night, I would walk through the stables and seek relief among my beloved animals."[19] Whereas Himmler mentioned animals to demonstrate the moral superiority of the Nazi regime, Hoess attempted to use them as proof of his sensitive, empathetic nature. However, the fact is he merely felt sorry for himself for what he had "had to . . . watch."

The stories about Koch's concern for zoo animals, Hoess's reliance on horses, and even Hitler's oft-cited weakness for German shepherds are, if nothing else, also part of the legend of the Nazis' purportedly advanced ideas about the importance of protecting animals and nature, which to a certain degree has held up through today. Even now, reference is still made to the fact that during the first year of his rule Hitler already had a new animal welfare act issued that was internationally considered to be progressive, and which remained in force in West Germany largely unchanged until 1972. This legislation, the first of its kind in the German Empire, was designed to protect animals for their own sake; it even earned the self-proclaimed animal lover Hitler a medal in the United States.[20] As early as August 1933, when he was Prussian minister-president, Hermann Goering had railed against animal experiments and threatened vivisectionists with the concentration camp—one of the first public mentions of the camps, incidentally. In this case, however, it went no further than empty threats.[21]

All of that only appears to be self-contradictory, for the protection of animals is closely linked to fundamental convictions of Nazi ideology. Maren Möhring numbers among the few historians who addressed the topic of animals during the Nazi era until recently. In a 2011 essay, for example, she investigated in detail how the relationship between humans and animals changed in Nazi Germany. As Möhring writes, Nazi thinking on animal welfare—paradoxical at first glance—cannot be explained as either a propaganda tool or a positive aspect of Nazi ideology detached from the rest of it. Rather, it was much more an "integral component of a new order for society on an ethnonationalist and racist foundation."[22] Put another way, it was an ideology that measures the value of life by what use it brings to the particular community living together harmoniously and does not distinguish between *humans* and *animals* but rather between life that is useful and that which is unworthy of living. Thus, the practice of according some animals special protection, and, in turn, declaring some humans to be "pests" and systematically exterminating them derived from the same ideological spirit.

That mentality is shown in an especially pointed way again by the commandant of Buchenwald: Koch, who was so concerned about the well-being of the zoo animals, had prisoners thrown into the bear den for his pleasure and watched as their flesh was torn to bits by the animals.[23] After the concentration camp was liberated, Leopold Reitter, a Buchenwald survivor, stated for the record: "Even in 1944, when massive starvation prevailed in the camp, the birds of prey, bears and monkeys got meat every day, which, it goes without saying, was taken from the prisoners' kitchen and thus eliminated from the prisoners' meals."[24] There are a great number of these kinds of reports. Outside the concentration camp, too, animals show up in numerous diaries, memoirs, letters, and everyday documents. Until recently, however, they have played at most the role of extras in the research on National Socialism. Although since the 1980s historians have explored innumerable areas of everyday life under Nazism, from fashion to sports to nutrition, handicrafts, and drug consumption, animals have, until now, seldom been the topic of discussion.

The reasons for this dearth of discussion are obvious: scholars researching Nazism, especially German scholars, have reservations about even touching on the subject "because there is a fear that focusing on animals would lead to trivializing the human victims," according to University of Kassel's Mieke Roscher, who currently holds the only professorship in human-animal studies in Germany.[25] Yet it is precisely because this seemingly harmless history of animals is so closely interwoven with the everyday, as well as with Nazi ideology, that it is so relevant. If nothing else, it shows how deeply dangerous ideas were anchored even in areas of life supposedly not affected by ideology and how profoundly these ideas could inform society: anyone taking a closer look at cats as pets in the 1930s and 1940s will get a glimpse into German living rooms and will also come immediately face to face with a *völkisch*/racist view of the world, one that pervaded the everyday. Sooner or later, anyone who delves into insects in the Nazi years will find themselves in the classrooms of German pupils and inevitably dealing with "poisonous pedagogy" and social Darwinism. And anyone who wants to find out what role domestic pigs played at this time will run into not only advertising posters from the food industry and early forms of the recycling economy but twisted outgrowths of Nazi ideology as well. The histories of animals run at cross-purposes to many well-known topics in research on Nazism and, as a result, provide a frequently different, mostly new, but never trivializing perspective on life in the Nazi era.

Tracking Animals

The Nazi terror was not obvious everywhere. In many places, brown-shirted everyday life had more of a gray-on-gray color scheme. Yet in all areas of its life, animals were of importance, as the following chapters show, each of which approaches a different facet of Nazism by way of one animal species. In chapter 1, with the aid of the dog and its wild ancestor, the wolf, we take a look at race theory, showing how tightly enmeshed aspects of everyday life and ideology, politics, and "science" were with one another. By way of the domestic pig, in chapter 2 I offer more than just a reading of the importance of *Nutztiere* (working, or utility, animals) in Nazi Germany; I also demonstrate how the pig as the most important provider of fat and meat for the *Volksernährung* (the nutrition of the nation) was central to Nazi efforts to create a state that was not dependent in any way *on other countries and to prove the worthiness of its own "arische Urkultur"* (primal Aryan culture). In chapter 4, I investigate the domestic cat, which above all other animals highlights the ambivalent feelings that house pets provoked. For some, cats were a "Jewish animal" that could not be tamed. Others praised them as mousers and "hygienic aides to *Volksgesundheit* (the health of the nation)." In this chapter we meet diverse cat owners, including philologist Victor Klemperer, who with his wife feared first for the life of their cat, Mujel, and then for their own.

Animals defined pedagogy and education during the 1930s and 1940s as well. In the example of silkworms and potato beetles, the subject of chapter 3, we see how even the youngest were being prepared for war and combat. Insects, as can be demonstrated by textbooks and children's books, were used additionally to explain to children what—and, in particular, who—counted as "pests," "vermin," or "parasites" in the Nazi sense.

There was not any single uniform Nazi ideology with respect to animals. Aspects of the Nazi worldview were combined arbitrarily, in this way and that, a mishmash evidenced in an exemplary fashion in different views among Nazis toward hunting, as I detail in chapter 5: whereas Hitler mocked hunters as "green freemasons," Hermann Goering, the *Reichsjägermeister* (Reich hunting master), famously could not get enough of trophy hunting. Central to this chapter is Raufbold, or Ruffian, the red stag whose statue adorns the cover of this book. Having fallen victim to Goering's obsession with trophy animals, Raufbold outlasted the ages, as well as the twelve years of the thousand-year Reich, in his bronze image form—just as the legacy of Goering's worldview has survived, informing the enterprise of hunting to this very day.

Finally, when it comes to the role of animals in the Third Reich, we must not overlook the fact that World War II, especially on the eastern front, would not have been possible without millions of horses. In chapter 6, we accompany Siegfried, the Trakehner warmblood stallion, who was with his rider during the invasion of the Soviet Union in the summer of 1941 and who trekked even further east after that, long after engines and machines had given up the ghost in the cold of the Russian winter. This chapter shows how complicated the horse's symbolic significance was for the world image of the Nazis—and how long the shadow cast by this symbol is in today's Germany, too.

Drawing Boundaries

In *Minima Moralia: Reflections from Damaged Life*, a collection of aphorisms and short essays, Theodor Adorno notes that the "indignation over cruelty diminishes in proportion as the victims are less like normal readers." Drawing on this idea, he reasons:

> Perhaps the social schematization of perception in anti-Semites is such that they do not see Jews as human beings at all. The constantly encountered assertion that savages, blacks, Japanese are like animals, monkeys for example, is key to the pogrom. The possibility of pogroms is decided in the moment when the gaze of the fatally wounded animal falls on a human being. The defiance with which he repels this gaze—"after all, it's only an animal"—reappears irresistibly in cruelties done to human beings, the perpetrators having again and again to reassure themselves that it is "only an animal," because they could never fully believe this even in animals.[26]

For Adorno, the relationship of human beings to one another was also reflected in their interaction with animals. In this sense, the history and the stories of animals in the Third Reich are not only testimonies of their age. Animals further reveal the image of humanity and of the world that this era brought forth and in the end, therefore, play much more than the role of mute extras.

I

Blood Ties

I have given a name to my pain and call it "dog."
    —Friedrich Nietzsche, *The Gay Science*

The stranger appeared as if out of nowhere, so suddenly that the dog did not even sense him at first. After all, he was totally preoccupied with hunting down a rat. Through trenches and concertina wire he had pursued it, and he must have somehow lost his direction and then ended up behind enemy lines without realizing it, where all at once there was this man standing before him. When the man reached for him, he bit down as hard as he could, but the man did not let go and instead dragged him away, into a dark space beneath the ground, where the air was damp and cool and smelled like humans.

The stranger was a twenty-three-year-old German lance corporal. On this day in the spring of 1915, he was as usual on the way to his unit, the 16th Royal Bavarian Reserve Infantry Regiment. They had pitched camp in the cellar of a castle in Fromelles, a village in the north of France, only a few kilometers from the western front. The soldier was amazed by the little deserter with the white coat and a black spot stretching across his left ear and eye. Since the dog came from the direction of the British position and resembled an English fox terrier, the soldier named him "Foxl."[1]

After reaching headquarters, the lance corporal tried to earn Foxl's trust with cookies and chocolate. The young man did not know all that much about dogs; otherwise he would have been aware that chocolate is poison to a dog and that, especially for such a small one, it can be deadly even in small amounts. The alkaloid theobromine contained in the cacao bean causes elevated blood pressure in dogs and constricts their blood vessels, which can lead to cramping, cardiac arrhythmia, and, ultimately, respiratory arrest. There was a war going on, though, and the chocolate the soldiers had in their provisions was of lesser quality, so—luckily for Foxl—it hardly contained any cacao. The dog soon overcame his initial shyness. He gradually got used to the quiet man, who then taught him how to jump over a rope and clamber up and down ladders.

The soldier's job was as a messenger; he was tasked with delivering messages from the regiment staff to the battalion staffs. "Etappenschwein" (roughly, "relay jackass") is what the soldiers disparagingly called those like him, since they were able to avoid enemy bullets and grenades because they moved around mostly in the back trenches. The epithet bothered the lance corporal so much that in later years he would always portray himself as a brave soldier on the front lines, doing everything he could to eliminate any doubts about his heroism.

The regiment was relocated multiple times, and every time Foxl went along, becoming something of a little mascot, one that the men liked to have their pictures taken with. Whenever the lance corporal was called out, Foxl stayed back at headquarters, tied up on his leash, and waited for him to return again. Starting in the fall of 1916, though, the lance corporal began staying away longer than usual.

At the beginning of October, the soldiers took up a position between the towns of Bapaume and Le Barque, around fifteen kilometers north of the Somme. There, ever since July 1916, British, French, and German units had been trying to grind one another down in a battle of attrition. When the fighting ended four and a half months later, more than a million soldiers had fallen on both sides. It was one of the most devastating battles of the war. Foxl's soldier was lucky and survived; after only three days in combat, shrapnel from a grenade bored into his upper left thigh, granting him around two months' leave in the military hospital in Beelitz, near Berlin.[2]

After he returned to his regiment in March 1917, his comrades reported that the dog would not listen to anybody and would hardly allow anyone to pet him. It filled him with pride to know that Foxl would only obey him because, besides this dog, there was no one else who would do that. Among his comrades, the man was considered to be a loner, an oddball who had no use for their locker-room antics and boasts, a weirdo who preferred to crawl away and hide behind his newspaper and his drawings or play with his dog. Foxl seemed to be the only living creature he was attached to.[3]

For many soldiers who went to war, dogs became important companions and close confidants. Whether it was the fact that they were the first to smell when poison gas crept into the trenches, thus warning the men in time, that they brought messages to the very head of the front lines, that they tracked down the wounded between the lines in no-man's-land, or that they comforted the soldiers, to many, dogs appeared to be messengers from a better world, giving them more hope than all the letters sent from home to cheer them up and all the truisms from superiors about keeping a stiff upper lip.[4] In a firsthand account of the experience on the Italian front during World War I, Robert Hohlbaum, a soldier in the Austro-Hungarian military, wrote that "in those days when the world was wobbling on the brink, the endearing bark of a dog gave more to us than the wisest words of any human."[5]

The young lance corporal could no longer imagine being without his Foxl. One day, when the war was over, if the dog were still alive, the corporal intended to provide him with a mate. Now, though, they were still somewhere in the hinterlands on the western front, sharing their feed and a folding cot. Two and a half years went by, with Foxl accompanying him on his march through the north of France and Belgium. Sure, he still hunted down the odd rat now and again, but in the end he always returned to the soldier. Until that one day in August 1917.

Just as the regiment was to be relocated once again, going by train in the direction of Alsace, it suddenly seemed as if Foxl had been swallowed up by the ground itself. The soldier had his suspicions. A short time prior, a railroad worker had offered him two hundred marks for the dog, but the soldier rejected the offer indignantly: "Even if you gave me two hundred thousand, sir, I would not give him to you!"

The soldier was certain that this *Schweinehund* ("bastard"; literally, "pig-dog") stole him. But there was no time left for him to look for Foxl; his unit was already on the move, with a long march on foot ahead of it. Feeling like he had lost his most faithful companion and not being able to do a thing about it, he got a move on, too, and tried to put Foxl behind him. No victim of any war would affect him as deeply as this loss of his dog.

Hitler's Hounds

What became of Foxl, Adolf Hitler would never learn. Hitler himself, as we know, returned to his chosen home of Munich, where, after a short period of being politically rudderless, he became radicalized and, in 1920, cofounded the Nationalsozialistische Deutsche Arbeiterpartei (National Socialist German Workers' Party). After the failed attempt at a putsch

against the Bavarian government in November 1923, he landed in prison
for several months. Subsequently, underestimated by his opponents yet
undergirded by willing supporters, he succeeded in climbing so high that
it was no longer just a dog running after him but, instead, nearly an entire
Volk (nation).[6]

At the close of January 1942, Adolf Hitler was sitting, as he so frequently
did at night, with his followers in Wolfsschanze (Wolf's Lair), the führer's
headquarters, situated in deepest East Prussia, telling of the loss of his first
dog, Foxl.[7] At that time the first forced laborers from the Soviet Union
were being dragged off to Germany. Of the estimated five million or so
Russian Jews, five hundred thousand had already been shot dead by the
so-called *Einsatzgruppen* (literally, "deployment groups," i.e., death squads)
from the Reichssicherheitshauptamt (Reich Main Security Office).[8] Just
a few days earlier, on January 20, fifteen high-ranking members from various government offices and the SS had met without Hitler around noon
in a villa by the Großer Wannsee. At this lake in southwestern Berlin, they
gathered for a "conference with accompanying breakfast," as it was described in the invitation. On the agenda that day, though, before brunch,
were "questions connected to the final solution of the Jewish question." The
Nazi bureaucrats had apprehended eleven million people across Europe.[9]
The mass murder of them was already in full swing.

Even if Hitler was not in attendance at this meeting about the genocide,
he was very well informed about all the steps that were going to be taken
and had given his imprimatur to them.[10] He left no doubt that the steps
introduced for expelling and exterminating European Jews were in line
with what he wanted: "Let them go to Russia. Where the Jews are concerned, I'm devoid of all sense of pity," he said, around a week after the
conference, during a further round of nighttime talks at Wolfsschanze.[11]
Here in this familiar atmosphere, where the war was far away and nobody
interrupted him, he liked talking the most about his feelings, telling of
Foxl, of Muck and Blondi, and of all the other dogs he had ever possessed.

He had had countless hounds already; how many it would be in the
end cannot be said with certainty, for the sources contradict one another on
this point. Adding to the difficulty of confirming how many dogs he owned
is the fact that Hitler gave the same names to multiple dogs—he called
at least three stud dogs Wolf and three bitches Blondi. But we do know
that from 1922 to 1945, he owned at least thirteen, all of them German

shepherds except for one. We also know that at one point, he had three dogs at the same time. On top of the dogs he kept, moreover, there were numerous other dogs that he gave to party members and fellow travelers on special occasions.[12]

Hitler described himself as an animal lover, though it is still a question whether he was a dog lover in the strict sense.[13] "Dog" was one of the most frequent curse words he used, regardless of whether he was raging during one of his nighttime monologues or threatening his enemies. In 1923, he had announced in the *Völkischer Beobachter*, the party organ of the Nazis, that he would "rather be a dead Achilles than a living dog."[14] On top of that, he made distinctions among dogs. He placed particular value on purebreds, even if he fundamentally rejected certain breeds. For example, he did not like bulldogs and boxers. Nor did he approve of dachshunds, which were originally bred to hunt badgers and pursue them into their dens—he was bothered precisely by the character trait typifying the breed, namely, their self-will. Hitler treasured dogs that did not have an overly strong will of their own, that obeyed an owner's commands and were compliant. Of all the *Hunderassen* (dog breeds, the German term connecting via "rassen" to the construct of race), he was particularly attracted to German shepherds.

At the same time that the Nazis declared the Jews, Sinti, and Roma to be subhuman and people with disabilities unworthy of living, they elevated several animals to the status of "master animal," including the dog.[15] Anything that did not fit into their hierarchy of values, anything that was what they described as alien to the race, degenerate, or sick, they tried to winnow away and wipe out.[16] How strongly these ideas were informed by their ideas about animal breeding is nowhere more clearly shown than in the case of the German shepherd.[17] This attitude did not just emerge in the 1930s, though. In order to understand it, it is worth taking a look further back in the history of the relationship between humans and dogs.

The *Rittmeister's* Race

It is unclear precisely when, but sometime between forty thousand and fifteen thousand years ago, humans and wolves got together. Presumably, human leftovers and garbage first attracted the less timid among these wolves to human habitation sites. It was probably the wolves that initiated the relationship, even if there were isolated cases of people adopting and raising orphaned wolf pups.[18] Because both wolves and humans live in close family groups, it was not all that difficult for either of them to adjust to the other. Over the course of time, the tamed wolves became dogs, and the once nomadic hunters became sedentary settlers. Several dog breeds are correspondingly ancient: by early antiquity, Portuguese water dogs were already helping seafaring peoples catch fish on the Atlantic coast, and St. Bernards and collies date to the Middle Ages.[19]

In contrast, the German shepherd's lineage is downright young, not nearly as ancient as its fanatic admirers might think. It was only in 1901, when the twelve-year-old Adolf Hitler was still attending middle school in Linz, Austria, that a certain Max von Stephanitz first established the traits for this breed, in his work titled *Der Deutsche Schäferhund in Wort und Bild* (*The German Shepherd Dog in Word and Image*).

Von Stephanitz himself had come across the dog by accident only a few years before. He had been interested in animals early on, to be sure, and had wanted to become a farmer, in fact. Then he gave in to his mother's wishes and embarked on a military career as a cavalryman, eventually achieving the rank of *Rittmeister* (riding master). Apart from a parrot at home, army horses were the only contact he had with animals for a long time, until one life-changing experience in the mid-1890s. During a combat exercise, he observed a shepherd who was rounding up his flock with the help of his dogs. He looked on in astonishment at the way the shepherd directed his dogs only with finger signs and calls. The interplay between human and animal reminded him of a military maneuver and fascinated him so much that he decided then and there to dedicate himself to what became known as the German shepherd.[20] What is more, he wanted to breed a canine race that would unite in itself the virtues of the Prussian soldier: loyalty, courage, perseverance, industry, and obedience.[21] The breed was to look as "wolflike" as possible and thereby approximate what von Stephanitz and his contemporaries imagined to be the Germanic *Urhund* (primal dog).[22]

In the various provinces of the Wilheminian Empire around the turn of the nineteenth century, there were of course multiple stocks of herding dogs. While they strongly differed in coat color, physical build, and size, they closely resembled one another with respect to certain other traits. However, they no longer presented a uniform image of their breed. Even so, von Stephanitz was convinced that these dogs were all "members of one common, widely-diffused race," the origins of which, in his opinion, reached back to the Bronze Age.[23] By selecting the right breeding animals and culling everything that was pathological, he hoped to uncover this race again.[24]

To this end, von Stephanitz bought himself a bitch in 1897. The following year, at a breeder in Frankfurt am Main, he found a big, three-year-old stud dog. It stood a good sixty centimeters high at its withers, corresponding precisely to his expectations: "with powerful bones, beautiful lines, and a nobly formed head" and "clean and sinewy in build," the "entire dog was one live wire," von Stephanitz would rhapsodize. He revealed in this description of the dog's "essence" his very typical understanding of power for that era: the dog was "marvellous in his obedient fidelity to his master" and displayed "the straightforward nature of a gentleman with a boundless and irresistible zest for living."[25]

So as to bestow the purportedly select quadripeds with names befitting their standing, he called the bitch Freya, after the Nordic goddess of love and marriage, and he named the stud dog Horand, after a hero in Germanic sagas. As an addition, they got the name of his Upper Bavarian country estate of Grafrath, too—thus making their canine nobility a fait accompli.

A good year later in April 1899, when von Stephanitz cofounded the Association for German Shepherds (the still extant Verein für Deutsche Schäferhunde), Horand earned the stud book number 1 and became the progenitor of numerous generations in the German shepherd family tree.[26]

Von Stephanitz and his club colleagues not only wanted uniformity among the dogs but a specific combination of traits, and so they set their sights on two dog stocks above all others: small, compact animals from Thuringia and large, powerful ones from Württemberg.[27] Von Stephanitz thought nothing of so-called *Luxushunde* (luxury dogs) that were bred only for their appearance—above all, the German shepherd was to be a *Gebrauchshund* (utility, or working, dog). Not all breeders met von Stephanitz's expectations. Some of them crossed shepherds with wolves to prevent diseases like canine distemper or to achieve a more attractive, more "wolflike" physical build.[28] Von Stephanitz fundamentally rejected those methods for, in his opinion, they would produce only "shy, and for that reason, snappy dogs."[29] In order to justify his repudiation of them, he invoked a terribly abbreviated and exaggerated statement from Charles Darwin, according to which crossbreeding supposedly extinguished the "virtues of both parent races" and produced bastards lacking any character.[30]

Besides Darwin, in his work von Stephanitz referred above all to Ernst Haeckel, a zoologist and medical practitioner. Haeckel was one of the first to apply Darwin's theories on heredity to humans and was considered to be a trailblazer in the field of eugenics. The term "eugenics" had been coined by the British natural scientist Francis Galton—a cousin of Darwin's—in 1883 by which he referred to an improvement in human hereditary makeup through targeted and directed breeding. He was convinced that intelligence was exclusively hereditary and not conditioned by environmental influences and that therefore the intellectual differences among the "human races" (*Menschenrassen* in German, analogous to the term for dog breeds) were predetermined by their hereditary makeup.[31] To guarantee the health of the entire nation, then, the reproduction of healthy, intelligent humans ought to be promoted, while that of "sick" and "lower" humans ought to be prohibited.[32] Haeckel tapped into Galton's thinking and also pointed to the ancient Spartans, who, he was convinced, owed their beauty, strength, and intellectual energy to "the ancient custom" of killing all weak and disabled infants immediately after their birth.[33]

Eugenics rapidly found adherents everywhere in the world. By 1907, the state of Indiana had enacted a law for the compulsory sterilization of disabled persons and criminals. From the 1920s onward, similar laws arose in fifteen states in the United States, as well as in several Scandinavian and Baltic countries.[34] At this time in Germany, eugenics was known primarily as *Rassenhygiene* (racial hygiene), an expression that goes back to the physician Alfred Ploetz.[35] In the care of the disabled, sick, and poor, Ploetz, a friend of Haeckel's, perceived the sort of "humane touchy-feely sappiness" that "only hinders or delays the effectiveness of natural selection."[36] Just like Galton, Ploetz was of the opinion that the state had to regulate the reproduction of the population.[37]

Although efforts of this sort had already been made during the Weimar Republic, it would still take until 1933 before compulsory sterilization of the disabled, mentally ill, and alcoholics was prescribed with the Reichsgesetz zur Verhütung erbkranken Nachwuchses (Law for the Prevention of Progeny with Hereditary Diseases).[38] Until that point, racial selective breeding remained restricted to animals even if von Stephanitz had, following the lead of Haeckel and Ploetz, by now taken matters a step further. Having, among other things, recommended bludgeoning to death misshapen, weak, or superfluous whelps, the dog breeder also suggests that "we can compare our shepherd dog breed without exaggeration with the Human Race."[39] In the "Bestimmungen über die Führung des Zuchtbuchs" (Stipulations Concerning the Keeping of the Stud Book), he had determined that the "portion of foreign blood" in the breeding of German shepherds was to be specified up to 1/128. Just like the German shepherd, the German *Volkskör-per* (body of the nation) should be purebred.[40] In 1935, the Nazis seized on von Stephanitz's "breeding stipulations" in introducing the Erste Verordnung zum Reichsbürgergesetz (First Decree on the Reich Citizenship Law, contained in the legislation also known as the Nuremberg Laws).[41] It was now no longer about the *Blutbeimischung* (blood admixture) in German shepherds but rather about the *Mischlingsgrad* (degree of mixed race) in Jews and Roma.[42]

At the time, von Stephanitz struck a nerve with his targeted breeding of the German shepherd. In the case of animals, being purebred was considered desirable in nineteenth-century bourgeois circles, because owners thought a purebred animal's behavior could be more precisely predicted than that of a mongrel.[43] Within twelve years, the number of animals recognized as purebred climbed from 250 to around 13,000.[44] By the beginning of the 1930s, the stud book comprised more than four hundred thousand German shepherds.[45]

Within a few decades, the German shepherd had thus been elevated to the hallmark German hound. While at the turn of the nineteenth century the Great Danes of Otto von Bismarck, chancellor of the German Empire, had been respectfully called *Reichshunde* (hounds of the Reich) and the dachshunds of Emperor Wilhelm II had been seen "as the symbol for the jovial German," from now on von Stephanitz's German shepherd would stand for Germanness par excellence.[46]

The Nazis also sought to capitalize on the breed's purportedly soldierly virtues in war. As the war began in 1939, a series of mobilization orders was issued to private dog owners. In the first year of the war alone, an estimated two hundred thousand dogs were conscripted from German households. Besides the German shepherd, Airedale terriers, Dobermans, Rottweilers, giant schnauzers, and boxers were especially sought after. The role of dogs in the military was clearly different from what it had been during World War I: the era of dogs as medics and messengers was past; now, in fact, it was the large breeds that were needed, especially as tracking dogs and guard dogs. Because the losses on the eastern front were particularly high, the requirements were relaxed in 1941, permitting even mixed dogs—often designated disparagingly as "bastard dogs"—to be mobilized, so long as they were more than fifty centimeters high at their withers.[47] The Nazis quickly adapted their ideology to these new realities, as they so frequently did in other arenas.

One of the German shepherd's greatest admirers was Hitler. Once he had received his first German shepherd as a gift from his fellow party members for his thirty-third birthday on April 20, 1922, he could not imagine having any other dog at his side. Being inclined to romantic kitsch, Hitler named the dog Wolf.[48] For he also had a certain predilection for the dog's wild primal ancestor. In familiar circles, he even liked to be addressed as Wolf; he signed his private letters with this nickname and, right at the beginning of his political career, he took his lodgings in hotels under the name Mr. Wolf. On top of that, several of his numerous headquarters bore "wolf" in their names—like Wolfsschanze in East Prussia, Wolfsschlucht (Wolf's Gorge) in occupied Belgium, and Werwolf (Werewolf) in Ukraine.

That term "werewolf" would acquire yet another meaning by the end of the war. Starting in 1944, SS Chief Heinrich Himmler used it to refer his paramilitary units carrying out acts of sabotage against the advancing Allies in the German border regions. And in a radio address at the beginning of April 1945, Minister of Propaganda Joseph Goebbels incited the German population to violent resistance with the words "The werewolf himself is holding court and is going to decide over life and death."[49]

Mortal Enemy / Fetish Animal

"There is nobody the dog hates as much as the wolf," wrote cat owner and dog hater Kurt Tucholsky in his biting *Traktat über den Hund* (Treatise on the dog) at the end of the 1920s. For "the wolf reminds him of his betrayal in having sold himself to humanity—hence he envies the wolf his freedom."[50]

The fact that a love for dogs and a cult attachment to the wolf do not necessarily contradict each other, however, is pointedly proven by the Nazis; among all the animal symbols of their ideology, the wolf was used the most. And because they preferred to see themselves as predators, it is not surprising that in April 1928, a few weeks before the German parliamentary elections, their top propagandist Joseph Goebbels threatened the political competition with the words "We come as enemies! Like the wolf that bursts in on the flock of sheep, that is how we are coming."[51]

Wild animals, and particularly beasts of prey like the wolf, enjoyed a higher ranking in Nazi ideology than domestic animal breeds because domestic breeds had supposedly become effete. Scientists like the Austrian behaviorologist Konrad Lorenz supported this thesis. Lorenz, who joined the Nazi party shortly after Austria's annexation to the so-called *Altreich* (the German Reich as it existed territorially in 1937), was convinced that among civilized humans "all the physical and moral symptoms of decline are in essence the same as the symptoms of domestication in house pets."[52]

For the Nazis, the wolf was not only propaganda material but also a part of a Germanic *Urwildnis* (primal wilderness) that they imagined and that they believed needed to be rejuvenated. For that reason, at the 1937 International Hunting Exhibition in Berlin, visitors were shown 129 dead wolf trophies in a special exhibition called *Urwild* in addition to replicas of eradicated aurochs and wisent. For like these powerful wild bovines, the wolf had disappeared from German forests long ago.

For centuries the wolf had been feared, hated, and hunted. In the nineteenth century, even the so-called father of animals, Alfred Brehm, described it as a "four-legged robber and murderer," one that inflicted "monstrous damage," supposedly following legions "in a time of perpetual wars."[53] That was why in 1814, Friedrich Wilhelm III, the Prussian king, legally obligated the crop farmers and cattle owners of his realm to hunt wolves.[54] The farmers proved to be compliant subjects: more than a thousand animals were shot over many years, and by the second half of the nineteenth century, the wolf was close to being exterminated in German territory.[55] In 1904, the last wolf was shot to death in the region of Lusatia in the east; in the west, the last four remaining animals were slain in Alsace in 1911.[56]

When the Nazis finally seized power, there were only a few isolated animals roaming around the forests of East Prussia.[57] Even in those parts, in spite of its special significance for Nazi ideology, the flesh-and-blood wolf hardly had any chance of surviving after 1933. Though it did count as "nichtjagdbares Haarwild" (furred game that cannot be hunted) according to the Reichsjagdgesetz (Reich Hunting Law), it was not by any means afforded protection for that reason, as the literature on the subject has noted time and again.[58] On the Rominter Heath in East Prussia—where Hermann

Goering often went on the prowl in his state hunting reserve (and where we set out for in a later chapter)—wolves from Poland and the Baltic states would appear every so often. Every time, however, they were hunted with great zeal by Goering's *grüne Schergen* (green henchmen).[59] Thus despite strongly identifying with "wolfish" properties, the Nazis would knowingly forsake the protection of their "fetish animal," most likely out of consideration for an important part of their electoral base—farmers, who since forever had seen the wolf as their mortal enemy, always out to get their cattle.[60]

In Nazi hunting literature, wolves were praised for being "excellent medics" and "gamekeepers for red deer," only killing as much "as they need for nourishment."[61] Resettling them was even considered: "If it were possible to accustom the wolf to being fed, especially in winter, then it ought to be possible, perhaps even for a country with a high level of culture, to afford at least a small population of wolves," Wilhelm Bieger wrote in his 1940 *Handbuch der Deutschen Jagd* (Handbook of German hunting).[62] For Bieger, it was not so much about preserving the wolf as a wild animal. Rather, his sole aim was to resuscitate the wolf hunt, for it was supposedly "uncommonly appealing."

For the time being, the wolf would roam through homeland forests only in the brown shirts' ideology. It was the "wolflike" German shepherd that would have to serve as the pivotal propaganda animal instead.

Fräulein Braun's *Frondeuse*

A photograph of Hitler from the 1930s depicts an alpine idyll. Wearing a suit and Tyrolean hat and trying hard to look laid back, he lounges in a mountain meadow. With his right arm supporting his upper body, his left leg bent upward, he gazes past a panting German shepherd out into the distance. Down to the scenery and the panting dog, every detail was posed in this photo, meant to show a supposedly private side of the "führer-in-waiting."

The illustrated volume of a Hitler as nobody knew him was published in 1932, with a print run of four hundred thousand copies. Heinrich Hoffmann, Hitler's court photographer and *Reichsbildberichterstatter* (Reich press photographer), took the pictures. The alpine motif with German shepherd adorned the title page, showing that Hitler's alleged love for dogs was a fixed component of the führer myth from the very beginning.[63] Hitler, as is well known, wanted to be portrayed in specific ways. He did not, for example, want to be photographed wearing either shorts or spectacles. His desires in this regard extended to which dogs he would allow himself to be photographed with. No breed other than a German shepherd was ever photographed with him—neither his beloved Eva Braun's two black Scotties, Negus and Stasi, which he disparagingly called "hand brooms" (even though he played with them every now and then), nor the little terrier that lived at his Berghof compound and that, gritting their teeth, Hitler's court entourage called the "großdeutscher Reichshund" ("greater Germanic Reich hound"), on account of its slight size.[64]

In 1939, after his black German shepherd Muck and its mate Blondi died, the führer went without a dog for a few years.[65] That would change, though, at the beginning of 1942, when he acquired a female German shepherd. Where this particular German shepherd bitch came from is uncertain. One version purports that Martin Bormann, the leader of the party chancellery and one of Hitler's closest confidants, had been contemplating how he could cheer him up. Hitler's mood had suffered severely as a result of the setbacks on the eastern front during the winter of 1941–42, and Bormann had recently learned that the German shepherd of Gerdi Troost, the widow of Hitler's favorite architect, Paul Ludwig Troost, had had a litter the previous year.[66] And since Bormann noted every last one of Hitler's insignificant utterances for his own purposes, it had not escaped him that Hitler wanted most to have a German shepherd bitch.[67] Bormann, so the story goes, consequently picked out a little puppy with a conspicuously light coat, which Hitler named Blondi, the third dog he gave this name.[68]

A different version purports that three years after the death of Blondi the Second, Hitler purchased a new German shepherd bitch from a postal worker in Ingolstadt.[69] However he came to own the dog, what is certain is that Blondi the Third became Hitler's closest companion from then on. Whether at the Berghof in the Alps or in one of his headquarters, she was everywhere at his side, traveled in his private train, and was one of the few living creatures to enjoy the privilege of being allowed to ride along in his limousine.[70]

When the mood would take him, Hitler sang with her. At his command "Blondi, play schoolgirl!" she would sit herself next to him on her hind legs and put her front paws on the arm of his chair. He then prompted her with "Blondi, sing," encouraging her with a long drawn-out howl, which she then tuned in on. And if her howling sounded too high to him, he only needed to say, "Blondi, sing deeper, like Zarah Leander!" And, as his secretary Traudl Junge recalled, Blondi would obey.[71]

Blondi learned additional tricks, too; soon she could balance on a beam, clamber up ladders, and jump over a two-meter-high picket fence. The German newsreel *Wochenschau* tried to capture these scenes so as to show the German *Volk* how much their führer loved animals and how strong a leader he was.[72] How many of these tricks Hitler personally taught Blondi is questionable, though, for the dog spent most of its time with Fritz Tornow, a sergeant in the military, whom Hitler designated *Hundeführer* (dog führer, i.e., handler).[73]

The constant presence of Blondi displeased Eva Braun, who was downright jealous of the dog. She would occasionally give the German shepherd bitch a kick when she was lying under the table as Braun and Hitler dined, because it would make Blondi whimper, which in turn would lead Hitler to promptly admonish her to be quiet. Braun called it "her revenge," although it had no effect on Hitler's relationship with the dog.[74] Hitler's staff attested that he had a more intimate relationship with Blondi than with anyone else, including Braun. Albert Speer also thought that she played the biggest role in Hitler's private life and "meant more to [him] than [his] closest associates."[75] Even Goebbels noted in his diary that "the canine can do anything it wants in his bunker," after he had visited with Hitler in March 1942. "At present it is the object closest to the Fuehrer's heart."[76]

For this much affection on his part, Hitler expected absolute loyalty. In Hitler's orbit, most took care not to become too familiar with the dog because they worried about angering him should the dog show an interest in them.[77] He would rage with jealousy at anyone whom Blondi "trustingly nuzzle[d] up to," even if it were a high-ranking visitor like Ferdinand Sauerbruch, the surgeon and general physician for the German army.[78] In the summer of 1942, Sauerbruch was invited to the führer's headquarters. When the door in the office he was waiting in was opened to admit him, it was not Hitler but Blondi who came storming into the room, jumping up on Sauerbruch, barking and baring her teeth. Sauerbruch knew his way around dogs. It was on dogs that he had even undertaken his first attempts at operating on the lung of a living patient. Sauerbruch did not move but just quietly and persuasively talked to the dog until she calmed down and let him be. When Hitler eventually stepped into the room, his Blondi, who otherwise let nobody hold her, was sitting nicely next to the guest, giving him her paw. "What have you done to my dog, sir?" Hitler screamed at Sauerbruch. "You have lured away the only creature that is really loyal to me." In his rage, as Sauerbruch later wrote, Hitler let loose: "I'll have the dog shot to death!"[79]

The way the media portrayed the relationship between Hitler and his dog Blondi was a fixed component of Nazi propaganda. Yet the German shepherd also stood for another side of the Third Reich, one there are no scenes of in the Nazi newsreels. For these moments did not play out in the mountain idyll of the Alps or in the clearings of the woods surrounding the führer's headquarters.

"Ravenous Beasts"

Oswald Pohl faced an organizational problem. Since April 1942, he had been running the SS Wirtschafts- und Verwaltungshauptamt (Main Economic and Administrative Office) and so the German concentration camps also fell within his range of duties. At this time, more and more detainees from the conquered regions in Eastern Europe were being put into camps, now serving as *Arbeitskräftereservoirs* (workforce reserves), as they were officially called. Often the prisoners had to work outside the camp in quarries, sandpits. or coal mines. Owing to deployment to the front lines, however, Pohl did not have enough personnel at the camps to watch them at all times.[80]

On April 30, 1942, therefore, Pohl told all the camp commandants in a memorandum that the surveillance system had to become more flexible, namely, by way of "guards riding between posts" and, in particular, by "deploying guard dogs." By this point, the military and police had been using them for a long time. In 1937, Pohl's superior, Heinrich Himmler, had built the population of police dogs up. At his directive, beginning in June 1942, Pohl set up a dog-training squad of his own at the Sachsenhausen concentration camp north of Berlin.[81] From then on, concentration camps without dogs were inconceivable. And they would not only serve as guard dogs at the camps.

Almost everyone in Treblinka knew Barry. Sometime at the end of 1942 or the start of 1943, he had arrived there, though nobody knew exactly when. Treblinka, located around seventy-five kilometers northeast of Warsaw, was his third camp. Before that he had been in Trawniki and then in Sobibór. He was not a detainee, though. Barry was a St. Bernard mix—and as large as a calf. His black-and-white spotted coat did kind of remind people of a Holstein cow. Perhaps his name alluded to that famous avalanche dog that, in the fourteen years he lived, supposedly saved the lives of more than forty persons in the Swiss Alps.[82] Barry was basically a good-natured animal. He might be found lazing around in the sun or straying after the fox, deer, and birds in the camp zoo located next to the guards' residential barracks. He also got along well with the detainees and allowed them to pet him—in principle, that is, and as long as he was alone. For the most part, though, he accompanied his owner, Kurt Franz. Then Barry showed his other side.

Franz, twenty-eight years old, came from Ratingen, in the vicinity of Düsseldorf, which could be clearly heard in his sing-song Rhineland accent. On account of his red cheeks and his flawless appearance, the prisoners called him "lalka"—the babydoll.[83] Yet the harmless nature of the nickname was deceptive; Franz was Treblinka's deputy camp warden. Treblinka was the third extermination camp after Belzec and Sobibor that was created in the wake of Aktion Reinhardt (Operation Reinhard).[84] Hiding behind this code name was the plan to murder all the Jews, Sinti, and Roma in occupied Poland, as was once again confirmed at the Wannsee Conference in January 1942. In July 1942, Treblinka had become operational. Over the next thirteen months, just short of one million people were taken there. Most of them survived only a few hours, before they were shot to death or gassed.[85]

For Kurt Franz, the detainees were not humans. That was also what he said to his dog whenever he took him on walks through the camp, on the lookout for "die Gestempelten" (the stamped). That was what Franz called those who were marked as a result of being transported or tortured.[86] When Franz came across an inmate during his inspection rounds who did not behave in accordance with his expectations, he only had to say to Barry: "Barry, my man, grab this dog!" And just like that, the dog would bite down on the detainee.[87]

After the Germans were defeated in Stalingrad in February 1943, they were forced to retreat, and the Red Army gradually pushed the German troops back westward. Deportations also ground to a halt. There was no stopping the front now; it moved nearer to the occupied areas and so closed in on the camps as well. Himmler issued commands intended to improve the security of the camps and thereby prevent uprisings. He had their outer limits fortified and even partially mined. On top of that, he insisted that the SS guard details become more efficient, so the dog squads were assigned a new task: "Dogs that work the grounds outside the camp must be raised to be the kind of ravenous beasts that the Cape hunting dogs in Africa are," wrote Himmler in February 1943. "They must be trained in such a way that, with the exception of their keeper, they will tear anyone else to bits. Accordingly, dogs must be kept apart so that no accident can occur. Only when it is dark and the camp is closed down can they be allowed out, and in the morning they must be penned up again."[88]

The way Himmler imagined it, these dogs would round up and encircle the detainees like a flock of sheep, as he told Rudolf Hoess, the commandant of Auschwitz, during a visit. Himmler hoped to cut back on guard positions by using dogs in this way. He was convinced that if his plan worked, then one guard position with several dogs could monitor up to a hundred prisoners.[89]

Everything was precisely organized: what dog breeds were especially appropriate, how large their kennels should be, and how much they got to eat.[90] The predetermined 150-gram daily amount of feed corresponded to just a small portion of the nutrition a fully grown German shepherd needs. It was knowingly kept low so as to keep the animals always hungry and potentially aggressive.[91]

Airedale terriers, Dobermans, and boxers were also conscripted.[92] Yet it would soon become clear that not all breeds were equally suited to be guard dogs and tracking dogs. Several camps reported that boxers, in fact, did not obey as desired. On top of that, because of their short snout, they were less able to pick up a scent than other dogs. In July 1944, based on that poorer performance, all camp commandants were informed that "effective immediately, all boxer bitches found among the service dogs [may be] mated only to Airedale terrier and German shepherd stud dogs."[93] Absolute obedience still took precedence over the dogma of being purebred.

Even so, as a letter to the camp commandants from the SS Wirtschafts- und Verwaltungshauptamt put it in March 1944, many dog handlers were either poorly educated or lacked "sufficient intelligence" or else did not have a "real interest in or genuine love for animals."[94] They exploited the fact that they had nothing else to do but take care of their animals and guard the prisoners. Most of them lazed about, amused themselves with other keepers, played with their dogs, or set them on the prisoners to pass the time—not that they were punished for it, though.[95] For example, at

Plaszow, a camp near Cracow, Commandant Amon Göth reigned like a sadistic seigneur, regularly letting his dogs Ralf and Rolf loose on the prisoners.[96] At Austria's Ebensee, a satellite of the Mauthausen concentration camp, the SS guard detail deployed a Great Dane named Lord to hunt humans and torture detainees.[97] And Franz Bauer, the police officer stationed in the ghetto of Międzyrzec-Podlaski near Lublin, was feared among the internees there as the "henchman with the dog."[98] In all these places, the dogs served as symbols of power and oppression.

Yet it was not only male guards and supervisors who spread fear and terror with their dogs. At Ravensbrück, the largest concentration camp for women, the number of guard dogs was especially high, because Himmler believed that women were likely to be more readily intimidated than men by them.[99] It is said that the dog handlers there were supposedly even more brutal than their male colleagues and that they let the dogs loose on their prisoners even more frequently.[100]

In the summer of 1943, Himmler's spring worries came to pass: the six hundred remaining work prisoners in Treblinka succeeded in getting their hands on arms and ammunition, and they had also managed to hide several hand grenades in the zoo's dovecote. On August 2, 1943, shortly before four in the afternoon, they attacked the guards. They set the barracks on fire with Molotov cocktails and blew a fuel tank sky-high. The cloud of smoke could be seen from kilometers away. It was the first armed uprising in any SS concentration camp. Several hundred detainees were able to break out, though the dogs soon found their trail and quickly caught up to them. Most of them were apprehended and subsequently shot to death. A few, though, were able to escape into the surrounding forests. In the end, fewer than fifty persons would survive Treblinka.[101]

After this uprising, the Nazis began dismantling the concentration camp. Kurt Franz took over its direction for the last three months. The SS demolished the fences, barracks, and gas chambers, leveling the premises. A farmhouse was erected on the site of the camp, and lupines were planted all around, the idea being to create the impression that there had never been anything in Treblinka other than fertile cropland and forest.[102]

Franz was relocated to Italy soon thereafter to hunt down partisans and Jews, but Barry did not go with him; instead, he went to an acquaintance of Franz's, Major Friedrich Struwe of the medical corps, who ran the reserve military hospital in neighboring Ostrów. There Barry, whom everybody now just called "the big calf," behaved completely inconspicuously. He spent most of his time lying lazily under Struwe's desk. Barry did accompany Strewe on body searches. Yet Barry did not attack a single one of the hundreds of people who were lined up and standing next to one another naked, waiting to be inspected by Struwe. In 1944, Barry went to Struwe's wife in Schleswig-Holstein, where he survived the end of the war. Later on, Struwe's brother took over, until Barry was put to sleep in 1947 on account of old age. In all those years, he supposedly never bit anyone.[103]

When Franz and other SS men were made to answer for the mass murders at Treblinka at the district court in Düsseldorf in 1964, the court inquired about Barry's role. While Franz disputed the claim that he had ever set his dog on people, several witnesses, as well as a fellow defendant, said he had. The court then commissioned an outside evaluator to examine whether it was possible that Barry was only that aggressive in the presence of Franz. The appointed expert was none other than Konrad Lorenz, who had been running the Max Planck Institute for Behavioral Physiology in the Upper Bavarian town of Seewiesen since 1961. Lorenz, who had joined the Nazi party in 1938, boasted of having converted numerous students to Nazi ideology and had willingly Nazified his scientific writing.[104]

During the trial, Lorenz said that a dog is "the mirror image of the subconscious of its master." In the case of mixed dogs such as Barry, this reflection was supposedly even more pronounced because they reacted "much more sensitively than purebred animals." According to Lorenz, it was "recognized in behavioral psychology that the same dog can at times be good and harmless and at times dangerous and bitey, too. . . . It adapts wholly to the mood and temper of its master. If a dog enters into a new dog/master bond, then its character can even change completely."[105]

In September 1965, Kurt Franz was condemned to lifelong imprisonment "on account of the joint murder of at least 300,000 individuals and on account of murder in thirty-five instances of at least 139 individuals." In its verdict, the court described the case as one of "satanic cruelty."[106]

The merits of mixed-breed dogs, so praised by Lorenz, sound in retrospect like a peculiar contradiction in history. The German shepherd would outlast the Third Reich, though, to be sure, it would never quite lose the bad reputation as Hitler's favorite and as a Nazi propaganda animal. In the postwar years, especially in West Germany, the erstwhile model breed of Rittmeister von Stephanitz with its svelte "wolflike" form acquired an increasingly more massive shape and a strongly sloping croup, which corresponded to the new aesthetic ideal in its *Rassezucht* (pedigree; literally, "race breeding").[107]

In the summer of 1993, Kurt Franz, who had been on day parole since the end of the 1970s, was released from prison on account of his advanced age and health problems. In 1998, at the age of eighty-four, he died in a home for senior citizens in Wuppertal.[108]

2

Digestive Affinities

They know too much to understand. The pig knows less, that's why it understands more. It understands the truth of the beating heart and the truth of the axe.

—Szczepan Twardoch, *Drach*

It must have been a singular disappointment for my great-grandmother. Even though the pig reached its snout out toward her, grunting with expectation, its floppy ears hanging forward and waggling—she just saw its puny physique. When she rubbed her hand flat across its flank, she did not feel any fat, only ribs. She headed out with her wheelbarrow day in and day out, checking at the area pubs for kitchen scraps to give it so that it would eventually get round and fat. No pig gets fat from kitchen scraps alone, though, as my great-grandmother knew all too well.

During the harvest of 1904, fate brought her into the world in a field in the Austro-Hungarian province of Carinthia. Later on, after she had moved to Germany's Ruhr region, destiny did not treat her much better. The "English disease," as they then called rickets—a consequence of the years of famine during World War I—had deformed her bones early on, and of her five children, only my grandmother and her younger sister reached adulthood; the others survived hours or days at most. She took all these twists of fate in stride, though not without griping. She had always been a "poor critter," as she would later say, working like a dog her whole life but still denied the happiness she had hoped for. Chain-smoking was her only vice.

She had to take care of the household and the animals on her own. While her husband slaved in the mines underground, she hired herself out as a day laborer. They lived in one of the mining company's little brick houses in the south of Essen. Not far away was Villa Hügel, the estate of the Krupps. When people spoke of the "Waffenschmiede des Reiches" ("anvil of the empire"), they not only meant the factories of those steel barons but also all of Essen: by the end of the 1930s, with its approximately 670,000 inhabitants, it was the largest industry town of the Revier, as locals called the coal-mining Ruhr region. The south of Essen, though, still had a very small-town character. Here, in fact, the region had remained an agglomeration of villages and farming communities that had only a few towns up until the beginning of the Industrial Age in the middle of the nineteenth century.

Like so many miners' families, my great-grandparents kept chickens and rabbits, which lived in the narrow strip of yard behind their house, plus the one pig whose pigsty was a lean-to. Presumably, the pig was younger, that is, not a piglet any longer but at thirty to forty kilos still nowhere near fully grown. In all likelihood it was a *veredeltes Landschwein* (a Landrace cross-bred pig). This breed made up more than two-thirds of the entire German pig population at that time. The animals were not only considered to be robust and resilient but were thought to mature quickly.[1]

Though it had been living with the family for not quite a year, the pig did not have a name. That would just unnecessarily complicate the affair. In the end, it was there to serve only one purpose—to be slaughtered and eaten. *Veredelte Landschweine* are considered to be ideal *Mehrzweckschweine* (multipurpose pigs) but best of all for making into sausage and bacon. A fully grown sow will ensure an entire winter's supply of meat.[2]

The domestic pig is the *Nutztier* (utility animal) par excellence. No other animal uses what it is fed so quickly and effectively, no other grows up as fast and gains as much weight in such a short time, and no other is as sexually mature and fertile at such a young age. During this era, more than five million of the eighteen million households in Germany kept pigs; almost two million households were entirely self-sufficient. In 1937 alone, thirty-four million pigs were slaughtered. Their meat met around two-thirds of the entire demand in Germany.[3]

As a rule, a fattening pig is slaughtered at eight or nine months, as was also the case for my great-grandmother's pig, though it probably had not put on much more weight by that time. Still, a skinny pig was better than none at all. At the *Schlachtfest* (ceremonial slaughter), all the relatives got together, everybody hoping they would get something to take back home. The pig obviously had an inkling of what it was in for; it squealed loudly, trying to escape the barnyard, but with a bolt shot between the eyes, it was abruptly silenced. After it gave one final twitch, it was slashed through its throat and a trough was quickly shoved underneath so as to collect the blood gushing out. Everything was processed, nothing wasted. Even so, when the *Schlachtfest* was finished and the relatives gone, there was hardly anything left over for my great-grandparents.

The family lacked the money for concentrated feed that was required to fatten a pig up, and they needed the few potatoes there were for their own survival. Soon enough the war would dictate their meal planning. Then slaughtering at home would no longer be possible without official permission. But from now on the pig pen would stay empty. Whenever my great-grandmother spoke of her pig in the years after, she mostly dwelled on what sad shape it had been in: "As skinny as a German shepherd," she would say, taking a drag from her cigarette.

Although this anecdote perhaps recalls a distant, preindustrial world, it is in fact a mere lifetime ago. In our current era, the pig appears to be more of an industrial product than a *Nutztier*. It is indeed difficult to imagine that up until the middle of the twentieth century, the pig was a fixed component of everyday life, dwelling among humans in the country as well as in the city. Still, pigs have not lost their significance—pork, then as now, is the most important type of meat in Germany. On average, every German consumes around fifty kilos of it per year. Nearly sixty million pigs are slaughtered annually; the majority of them are imported alive, specifically for slaughter. On top of that, no country exports as much pork as Germany—and yet, while there are approximately twenty-seven million swine living here, we hardly ever set eyes on a pig in full, from snout to tail. Most of them lead a shadow existence in anonymous fattening facilities. If we do ever see them alive, then it is in passing on the autobahn, when they try to stick their noses through the air slits of the livestock trailer trucks taking them to the slaughterhouse.[4]

The removal of pigs from our midst began around 150 years ago, but it was a dragged-out process. At the end of the 1930s, pigs could still be found almost everywhere—whether in the cramped miners' housing of Essen's Revier or in the meadows of far-off Pomerania.

Empire of Pigs

Pomerania was famous for its pigs. More than 1.5 million of them lived in the broad swath of land along the Baltic Sea from the Darß Peninsula in the west to the Piasnitz (now Piaśnica) River estuary in the east. And the *Deutsche Edelschweine* bred there in the town of Schöningen were among the most desired kind of pig. This breed is especially robust and fertile and was used as a *Fleischschwein* (meat pig) above all else.[5] Whereas animals for breeding regularly won prizes at the annual exhibitions, animals for fattening were raised as *Karbonadenschweine* for the Berlin market, in particular, who owed their name to the especially treasured way that part of the pig was prepared in German cooking.[6]

The lord and master of Schöningen pigs was Hans Schlange. He called himself Hans Schlange-Schöningen, Schöningen being the name of the family seat where he was born in 1886. Before taking over his parents' country estate, Hans had dropped out of the University of Greifswald, where he had been studying agricultural science, to fight in the cavalry in World War I, during which he was wounded three times. He also joined the Deutschnationale Volkspartei (German National People's Party), which was very popular among Pomerania's *Landvolk* (rural population).[7] Soon thereafter, he was representing the party with a seat in the Prussian provincial legislature and from 1924 on in the German national parliament as well.

Within the party—an agglomeration of conservatives, nationalists, monarchists, and anti-Semites—Schlange-Schöningen was at first a *Rechtsaußen* (an extreme right-winger), but by the end of the 1920s he had grown increasingly more moderate. In 1929, he left the Deutschnationale Volkspartei and joined the Christlich-Nationale Bauern- und Landvolkpartei (Christian-National Peasants' and Farmers' Party). For several months beginning in the fall of 1931, Schlange-Schöningen belonged to the minority government of the Weimar Republic's chancellor, Heinrich Brüning, even though he had at one time opposed the republic. There, as a Reich commissioner in Brüning's cabinet, he was responsible for a program called *Osthilfe* that was created to support the rural economy of Germany's war-torn provinces in the east. In that capacity, Schlange-Schöningen endeavored to buy up large unprofitable operations and thereby strengthen smaller and midsized operations. His efforts did not at all please several among the land-owning Junker class surrounding the Reich president Paul von Hindenburg. They badmouthed him as an "Agrarbolschewist" ("Bolshevik agrarian") and called for "stomping on the head of the snake from Schöningen."[8] In the spring of 1932, his political career would come to an end—for the time being.

Even though he himself had long represented the ideal of the German farmer as the "source for German rebirth," he still found the National Socialists suspect.[9] He considered their *völkisch* cult around "Blut und Boden" ("blood and soil") pretentious.[10] When the Nazis seized power, he withdrew to his estate, where he would subsequently devote himself to breeding livestock.[11]

The feudal estate of Schöningen comprised 750 hectares in total.[12] The meadows extended to the western banks of the Oder River, which empties into the Baltic Sea sixteen kilometers further to the north near Stettin (now Szczecin). In the 1930s, cows, sheep, and pigs still grazed on the gently sloping hillsides. The noise of the propaganda in Berlin was still far away. Here, other sounds continued to characterize everyday life: the lowing of the cows, the bleating of the sheep, the metallic clanging when the estate manager hit the plowshare with two little hammers, ringing in the workday or ending it. Yet the sound of the pigs was the most prevalent. It was even more shrill than the squeaking of the fully loaded freight cars from the narrow-gauge trains that passed by. It was like the screeching of a buzz saw that literally cut to the marrow. They were shrieking from hunger. Only when they were fed would the noise soften into satisfied grunting and munching.[13]

For Schlange-Schöningen, the overriding principle was to keep the pigs "as nasty and as natural as possible." From spring until fall, while the sows and their piglets lived in a simple half-timbered stall, the remaining herd stayed in a five-hectare paddock. This pen was at a remove from the farmyard. There, meadow plants like clover and small animals constituted the pigs' main source of nutrition. Only in winter would the animals go into the stall, where they were fattened up with beets, potatoes, barley, and bran.[14]

At the time the Schlange family took over the estate, sheep still dominated the Pomeranian landscape.[15] That changed in the second half of the nineteenth century with the growth of potato farming, which was best suited to Pomerania's sandy soils.[16] Since potatoes constitute the ideal way to fatten pigs, the region subsequently developed more and more into pig country.

In the 1930s alone, Pomerania supplied eight hundred thousand slaughter animals across Germany for the purpose of *Volksernährung* (nutrition for the people).[17] For Hans Schlange-Schöningen, pigs and potatoes were inextricably connected. Pigs, he noted, determined the price of potatoes—

when the price of potatoes increased, then the price of pork decreased, and vice versa.[18] Admittedly, this dependency became a risk in times of war. For humans and pigs are omnivores and thus compete for food. That competition had caused many conflicts in Germany, as in January 1915, for instance.

Even though World War I had only been going on for half a year at the time, there were already indications that German supplies would get tight. As a consequence of the British sea blockade, raw materials and basic foodstuffs could no longer be imported, so the Kaiserliches Statistisches Amt (Imperial Statistical Office) conducted a survey among farmers to determine the extent of their supplies. Concerned about requisitioning, the farmers described their stockpiles as being significantly smaller than they were. As a result, German agricultural scientists concluded that there would not be enough foodstuffs to feed twenty-five million pigs in addition to the sixty million people in Imperial Germany. Therefore, farmers would have to use potatoes for people instead and so would have to drastically reduce their pig stocks. Within two months, five million pigs were slaughtered. The price of pork plummeted. Yet the market could not absorb this quantity of pork all at once. Even before the meat could be sold, a large portion of it had rotted, whereupon its price then skyrocketed. Suddenly, pork was considered to be a rare commodity and was traded on the black market, so farmers preferred to use their stockpiled supplies of potatoes and grain as pig feed rather than sell their reserves at a bad price. All the while, the slaughter of pigs would continue. In 1916, the swine population had already sunk to seventeen million.

This decrease in the number of pigs, in turn, had an effect on crop farming; everywhere in the country there was a lack of manure, so harvest yields declined. In the rain-soaked fall of 1916, the few potatoes there were in the fields rotted away. The consequence was famine, costing the lives of around a quarter million people. In the last year of the war, only a scant six million pigs remained, a good three-quarters less than there had been before the war.[19] The agricultural experts' plan for drastically decreasing the pig population in the shortest amount of time, thereby ensuring that the *Volk* were fed, ended in a fiasco and went down in history as the "Professorenschlachtung" ("slaughter of the professors"), as leading academicians and politicians were made to answer for the disaster.

Around twenty years later, in his 1937 book *Der Schweinemord* (Murder of the pigs), Richard Walther Darré would reinterpret this government failure as an "Anglo-Jewish conspiracy" against the Germans. For Darré, the pigs symbolically stood for the German *Volk*, whom the Jews were out to get.[20] Anyone wanting to understand the significance of agriculture in the Third Reich cannot ignore Darré. Beginning in 1933, he served as *Reichsbauernführer* (Reich farmer leader) as well as *Reichsminister für Ernährung und Landwirtschaft* (Reich minister of food and agriculture), which was remarkable above all because—unlike, say, Hermann Goering, Heinrich Himmler, or Joseph Goebbels—Darré had joined Hitler's circle only three years prior, that is, quite late in the game. And that is not only what made his leadership role in the Nazi Party special.

Darré was born in 1895, the child of a German merchant in Buenos Aires, where he lived until the age of ten. His goal was to return one day to his home of Argentina and to live there as a gaucho. Even though he ended up remaining in Germany, he was always seen as an *Aussteiger* (exiled maverick), and he never lost his Spanish accent.[21]

Darré began attending a *Kolonialschule* (colonial college) in 1914 in the province of Hesse in Germany. That was where better society sent their sons to have them educated for a career as a *Kolonialwirt* (proprietor in the colonies). He soon interrupted his training, however, and went to the western front to serve as a volunteer in the war. After the war, it was only with some difficulty that he found his way back to bourgeois life. Although Germany had to surrender all its colonies after its defeat, Darré took up his colonist training again, yet for disciplinary reasons he dropped out of the school shortly after getting his *Vordiplom* (preliminary diploma).[22] He then began affiliating with extreme right-wing groups, studying at the university in Halle, in the province of Saxony-Anhalt, where he would delve intensively into the breeding and natural selection of animals. His book *Das Bauerntum als Lebensquell der nordischen Rasse* (The peasantry as life source of the Nordic race) appeared in 1929 and his *Neuadel aus Blut und Boden* (A new nobility of blood and soil) the following year. In both these works, Darré dwelled on "Reagrarisierung" (a "back-to-the-land" movement) for Germany as well on ideas for its racial renewal.[23]

His thinking was not entirely new. The glorification of the peasantry had a long tradition in Germany, reaching back to the nineteenth century. For the ethnographer Otto Ammon, peasants were the "Jungbrunnen der Menschheit" ("humanity's fountain of youth"), and despite his critique of civilization, even Oswald Spengler—on whose thought the Nazis based a fair share of their ideology—saw in peasants "the eternal man" and "the ever-flowing source of the blood."[24] The *völkisch* movement's conviction that only a return to farming, to a life that was *bodenständig* (rooted to the soil), would be able to bring about the rebirth of the German *Volk* was grounded in these ideas.[25]

And all these ideas fit flawlessly into Nazi ideology. In no time at all, Hitler would commission Darré to work out a party program to win over the rural population and farming community. And yet at that moment, Darré was not even a party member.[26] And how much of a role he played in turning Nazism into a mass movement is in fact disputed.[27] Even so, the electoral successes in the countryside did contribute decisively to Hitler's rise, which in turn also strengthened regard for Darré.[28] Within a short time he developed into one of the "most ardent ideologues of the movement," coining the concept of blood and soil.[29] For him there was a symbiosis between the German *Volk* and its *Lebensraum* (living space), between *Rasse* (race) and *Heimat* (homeland)—in short, in Darré's view, where a person came from and what the person was fed were inseparably connected to each other.

In the beginning, Darré found a close ally in Heinrich Himmler, who named him head of the SS Rasse- und Siedlungshauptamt (Race and Settlement Main Office) in 1931. Darré had worked for him previously as an advisor for propaganda. Himmler saw the SS as a "racial avant-garde," as an elite within the Nazi movement—and Darré saw to it that the "Blut und Boden" ideology found its way into Himmler's SS.[30] Darré and Himmler complemented each other optimally. While Himmler played the functionary, Darré served as his ideas man.[31]

They had several other things in common. In the 1920s, both had joined the Artaman League, an organization made up of *völkisch* settlers who saw their purpose to be to move back to the lands east of the Elbe River, cultivate it on their own, and drive out the Polish seasonal laborers employed there.[32] They were also connected by their penchant for Germanic mysticism, which Hitler would dismiss as "cultish claptrap."[33] Moreover, Himmler also had an academic degree in agricultural science. At the beginning of his career in the party, he and his wife had bought a compound near Munich, where they intended to raise chickens as a way of boosting the meager salary of two hundred Reichsmarks he received every month as a party staffer. But the hens did not lay as many eggs as they hoped for, and so from then on Himmler concentrated on his political career in Berlin, leaving his wife on her own with their daughter and fifty chickens in Bavaria.[34]

To an extent, Himmler still operated like a farmer as he built up the SS. He was firmly convinced that the knowledge that had been gained from breeding animals and plants could be applied to humans. Thus, every so often, Himmler would sit bent over his desk scrutinizing the passport photos of the SS applicants under a magnifying lens and winnow out all those who had big crooked noses, prominent cheekbones, or dark hair.[35] He saw himself as akin to a seed cultivator, as he explained in a 1935 lecture, and he aimed "to sift out the humans who, according to their external characteristics, were not considered to be of use in building up the SS."[36]

Like Himmler, Darré had no scruples about transferring the animal breeding methods he learned in his studies onto people.[37] Beginning with the barnyard, his plan was to renew the Aryan race.[38] In the process, the pig would play a special role. As a student, he had been interested in and carried out research on the domestication of the pig.[39] In 1927, he published his pathetic, thirty-five-page work titled *Das Schwein als Kriterium für nordische Völker und Semiten* (The pig as the criterion for Nordic peoples and Semites). In it he wrote, "The interpretation of no other sacrificial animal has been so disputed; no domesticated animal varies in this way between being fully rejected and supremely venerated."[40] Darré explained the racial difference between Aryans on the one hand and Jews and Muslims on the other by way of the domestication of the pig. For Darré, the pig was the "Leitrasse" ("leading breed"; literally, "race") for Nordic *Völker*, the one that induced their Germanic forefathers to settle down.[41] "Whereas the domestic pig clearly shows that the Nordic people must have been settlers, the Semites just as clearly prove their nomadism by their rejection of everything connected to the pig."[42] As a forest animal, the pig allegedly depended on deciduous forests and water and was therefore a "faunistic antipode of every desert climate."[43] As opposed to the Nordic crop farmers, Semitic nomads supposedly had to bend to the will of their livestock herds and kept moving with them as soon as any pasture was fully grazed. In his eyes, their lives were thus "parasitic."[44] The pig thus served as proof for Darré of the racial and cultural superiority of Nordic peoples to Jews and Muslims—it was by the pig that one purportedly recognized the Aryan. It was the nurturing nature of these mammals that served, if you will, as their "Ariernachweis" (the infamous Aryan certificate).

As we know today, however, the oldest archaeological finds for domesticated pigs come, in fact, from the Fertile Crescent, which stretched from the eastern Mediterranean coast along the Euphrates and Tigris rivers to the Persian Gulf and which is considered to be the cradle of crop farming and livestock breeding. The earliest findings are almost ten thousand years old and were found in present-day Turkey and Iraq.[45]

This discovery is surprising given that the pig is regarded as unclean in Islam as well as Judaism. Consuming its meat is forbidden, not least of all because the pig is considered to be a coprophagist as it rolls around in its own excrement. That is not any natural preference, however, but instead a consequence of being kept in poor conditions: since pigs are unable to sweat, they seek out proximity to bodies of water where they can take mud baths that help them cool down. Only if they have no other option remaining before they collapse will they wallow in their own muck.[46] Similarly, as the American anthropologist Marvin Harris writes in his book *Good to Eat,* "let them get hungry enough, and they'll even eat each other, a trait which they share with other omnivores, but most notably with their own masters."[47]

The first pigs probably came to Central Europe during the *Völkerwanderung* (migration period) from the fourth to the sixth centuries.[48] It was under Emperor Charlemagne that pig farming first attained greater significance. In the practice known as pannage, swine were driven in herds into the extensive deciduous forests, where they scoured the ground for acorns and beechnuts. In the process, however, they also buried many seeds and thus contributed decisively to the spread of oak trees. Not least of all because of this, historian Joachim Radkau counts pigs as "among the unsung heroes of environmental history." In the seventeenth century, the question of who was allowed to drive their pig herds into the forest for pannage was so contested that it resulted in outright "pig wars."[49]

For centuries, forest grazing was the prevalent method of fattening pigs. Gradually, however, the forests dwindled and with them traditional grazing areas for pigs. They were then integrated into agriculture, which was only producing just enough yields to feed people. It was only the triumphant advance of the potato, beginning in the eighteenth century, that enabled pig populations to expand.[50]

"Pork-Barrel" Projects

Something a large part of the population did not suspect was that since the mid-1930s, Germany had been preparing again for a war. Pigs—in particular, pig fat—again played a decisive role. Besides butter, lard was the most significant source of fat. Lard had the advantage over butter in that it could be produced independent of the global market; feed had to be imported to fatten cows, but pigs—at least theoretically—could be fed just with potatoes, greenstuffs, and food scraps. Thus in Germany, unlike in the rest of the world, high-fat pigs came back into demand. Whereas the United States and Great Britain were increasingly breeding lean pigs—because consumers there preferred vegetable fats—Germany was banking on fatty pigs and, in so doing, on animal fat it would produce on its own.[51]

For Darré, pigs were more than an idée fixe that emerged from his theories on race. They were a decisive asset for preventing renewed food security issues. The famine problems of World War I and the economic crisis at the end of the 1920s had shown him that when in doubt, one could not rely on the global economy.[52] Never again—of that he was convinced—could Germany depend on other countries to feed itself. "Total war is not only a question of deciding about weapons but, in the first place, also one of ensuring nutrition for the nation," Darré wrote in 1937.[53] For this reason, he said, Germany must again become an autarchy with a self-sufficient *Volk*. In future, it ought to satisfy its demand for fat solely from its domestic agriculture. In addition, harvest surpluses that would otherwise go rotten could be fed to pigs, thus helping them get even fatter.[54]

Like German peasants, pigs ought to be *bodenständig*, that is, completely fed from homeland soil. For this reason, in July 1932 in the journal *Deutsche Agrarpolitik* (German agricultural policy), which he himself had founded, Darré once again called for fattening pigs more with native field crops like potatoes and beets instead of industrial feed.[55] When the Nazis seized power, the sites for pig breeding were relocated. The area around the port cities of Bremen and Hamburg in northwestern Germany was originally the pig region par excellence—since it was relatively easy to get imported feed grain and fishmeal there—but after 1933, pig production increasingly shifted eastward, in the direction of Pomerania, for instance, where there were sufficiently large expanses of land on which to cultivate potatoes for fattening pigs.[56]

Darré's principles regarding fat might have worked in theory, but the reality of it looked different. Even in 1936, Germany still had to import 60 percent of its demand for fat.[57] Accordingly, in order that Germany might be independent of foreign countries once and for all, the Reich demanded that everything that could be utilized in any way whatsoever be returned to the raw material cycle. In the Reich, this mania for reutilization stopped at hardly anything: just as both used and scrap metal were collected and melted down, so, too, the kitchen scraps of eighteen million German households were to serve their purpose as pig feed.[58]

For that reason, in November 1936, the Ernährungshilfswerk (Food Aid Organization) was created as part of the *Vierjahresplan* (four-year plan) that was supposed to make Germany ready for war and economically independent by 1940.[59] The organization was an affiliate program of the National-sozialistische Volkswohlfahrt (National Socialist Peoples' Welfare Agency), which until that point had been concerned, above all, with helping families in need of aid, with running kindergartens, and with evacuating children to the countryside. Now, everywhere in the Reich, pig-fattening facilities were being created where the animals were to be fattened exclusively with collected kitchen scraps.[60] So that enough scraps were gathered, homes and businesses were obligated by police order "to make the kitchen and food scraps accruing to them available to the Food Aid Organization and to throw the scraps into the house collection bins set up for that purpose."[61]

In numerous kitchens in the Reich, there were special enamel crocks with the inscription "Küchenabfälle für das Ernährungshilfswerk" ("Kitchen Scraps for the Food Aid Organization") in addition to the house collection bins. Advertisements on posters and in newspapers called on citizens to "Kampf dem Verderb" ("Fight the Rot"). One of the posters showed a housewife being stopped by a hovering hog as she was preparing to put leftovers into a trash can. The caption read "So ist's nicht richtig!" ("That's not the right way!"), and underneath it said "Kampf dem Verderb . . . und Deine Küchenabfälle dem Ernährungshilfswerk" ("Fight the Rot . . . and Give Your Kitchen Scraps to the Food Aid Organization").

The walls of numerous kitchens also proudly displayed the so-called National Socialist Peoples' Welfare Agency piggies. They were cardboard signs in the shape of a pig, distributed to homes by agency employees going door-to-door. In *Sütterlinschrift*, an old-fashioned German hand-writing style, the signs listed what was good for pigs, including potato peels, vegetables, salad, fruit, and meat and fish scraps—and what was not, namely, chemicals, cleansers, spices, lemons, oranges, and banana peels.[62]

When the pigs in the fattening facilities were fat enough, then they were loaded onto trailer trucks and taken to the slaughterhouses. Even on their way to death, they advertised for trash collection: a jingle appeared on the sides of the trucks that read "Wenn Du Küchenabfall hast / her damit zur Schweinemast" ("Whenever you have some kitchen waste / give it to the pig-fattening place").

As authorized executor of the four-year plan, Hermann Goering intended to produce one million additional pigs every year this way. Even though pushed and promoted energetically, the undertaking was nevertheless far from being as productive as expected. These results also demonstrated what my great-grandmother and every other pig breeder already knew: scraps alone could keep a pig alive but did not suffice to fatten it up. The production numbers remained well below expectations; by 1940, the pig population in the agency pens had increased just to thirty-thousand animals. A large portion of the collected kitchen scraps rotted, since they could not be preserved in time. On top of that, the Food Aid Organization was underfinanced from the start. Over the course of the war, the situation became even more acute. By 1942, the organization was in the red by more than sixteen million Reichsmarks. In February 1944, this debt was offset, whereupon the Reichsfinanzministerium (Reich Finance Ministry) decided that the Food Aid Organization must cover its costs on its own going forward.[63]

Although mistakes made during World War I were supposed to not have been repeated, the supply situation in Germany worsened dramatically during this war, too. Added to that were unexpectedly cold, long winters that would lead to harvest failures.[64] That had consequences as well for Darré personally. By 1944, the once-influential *Reichsbauernführer* and *Reichsernährungsminister* was largely relieved of his powers. In the course of a few years, he had fallen out with numerous Nazi functionaries. Darré even fell out of Himmler's good graces, because he supposedly saw the German settlement question too much from the "perspective of food policy."[65] Darré's image of the German peasantry was limited to the soil of the homeland. Hitler and Himmler, however, meant to settle the *Lebensraum* in the east not with mere farmers but rather with *Wehrbauer* (soldier peasants).[66] They had in mind an eastern wall, made of living people.[67] It was for this reason that Himmler removed his former friend as head of the SS Rasse- und Siedlungshauptamt in 1938.

The first hint that Darré's career was going downhill came in 1936, when Goering was putting together his staff for the four-year plan and did not assign Darré to lead the Geschäftsgruppe Ernährung (Food Business Group). Instead, he went with Herbert Backe, Darré's second-in-command at the Reich Food Ministry. As regard for Darré decreased, Backe became increasingly powerful in his shadow. And as it became more and more clear that Germany would be unable to feed itself on its own during the war, Darré was made to answer for that incapacity. In 1942, he was given his leave both as *Reichsernährungsminister* and as head of the Reichsamt für Agrarpolitik (Reich Office for Agricultural Policy). Backe took over all Darré's important offices.[68] And so the time of ideologues in agriculture and food policy came to an end; it was now the technocrats who would reign. There had long been jokes making the rounds about Darré the agroromantic. One of them was an epitaph for a German pig: "I, too, suffered a hero's death / I died from Darré's barley scraps!"[69]

While the demoted Darré withdrew into private life, Hans Schlange-Schöningen was dealing with the effects of Nazi agricultural policy on his estate in Pomerania. The situation in the countryside was getting worse from one year to the next. Farmers like to complain about the weather, to be sure, but the weather during the war years was in a special class of its own. In 1941, the winter dragged on into April, and a hard freeze set in already that fall. It was so cold at the beginning of 1942 that the Rhine and the Oder River both froze over, the consequences of which were fatal. The harvests were meager; what was vitally essential became scarce. "Millions of hundredweights of potatoes and turnips have been ruined by frost," Schlange-Schöningen wrote in his notes, published one year after the war ended. "Our cattle-stocks are rapidly diminishing, and lack of fodder has led to a slaughter of our pigs far greater than took place during the First World War. Will it be possible to scrape through until the next harvest? Ideological propaganda fills no hungry belly."[70]

The problems Germany confronted during World War I thus reappeared. Because grain, in particular, was insufficient as a result of the failed harvests, potatoes went from pig feed to the main food source for people during the war.[71] Once again, pigs were seen as competing with humans for food. In many places, the farmers found themselves in a double bind: because they had to deliver a large share of the potatoes to feed the nation, there was not enough left over for their animals.[72] The *Zeitschrift für Schweinezucht* (Journal for pig breeding)—the official magazine for German pig farmers in the Reichsnährstand (the Reich Nutritional Professions, an agricultural cartel)—delivered the usual slogans that called for keeping a stiff upper lip ("The German pig farmer knows his first duty: to persevere and aid our victory by ensuring food for the people!") and that discouraged dissent ("Anybody using bread grain as feed is helping the enemy!").[73] In addition, more and more firsthand reports of farmers' experiences as they tried to also pull the pigs through were published. Beets and greenstuffs, in particular, were now on the menu, and farmers were also supposed to put their herds out to pasture again and, in the fall, drive them into the harvested fields so that they could rummage for leftover tubers, roots, and small animals there. What might almost sound like an idyllic organic farmyard to contemporary readers was nothing but a desperate attempt to manage the prevailing paucity, one way or another.

Moreover, the directives from the Reichsernährungsministerium were contradictory. On December 18, 1943, it ordered "the intensified removal of pigs"—that is, a mass slaughter—in those regions in which there were too few potatoes for both humans and animals.[74] Just one day later, the *Zeitschrift für Schweinezucht* published an appeal from Reichsernährungsminister Backe to "preserve the pig populations," for only in this way, it said, would one be able to avoid "any infringement on the essence of the farm to the disadvantage of the future."[75]

Forbidden Flesh

Even so, any unrestricted slaughter had been impossible for quite some time. Food rationing began before the war began. Having their own pigs and other *Nutztiere*, one-third of the population was considered to be self-sufficient at this point in time, yet an official permit for slaughtering was required in every case.[76] *Schwarzschlachtungen* (illicit—literally "black"— slaughters) were punished draconically: on September 4, 1939, three days after the invasion of Poland, the Kriegswirtschaftsverordnung (War Economy Ordinance) came into force. Concerning "conduct detrimental to the war," it warned that "whoever destroys, hides or withholds raw materials that are vitally essential to the needs of the people and who as a consequence

maliciously endangers supplying this demand will be punished with incar-
ceration in a correctional facility or prison. In especially grave cases the
death penalty can be imposed."[77]

In rural regions, though, *Schwarzschlachtungen* continued to be common practice. Admittedly, it could be difficult to keep this practice secret
in the case of every animal. Slaughtering a chicken or a rabbit without
being caught was relatively easy. But it was harder with a pig weighing
100 to 150 kilos. Frequently, a good portion of the animal slaughtered went
to the neighbors for keeping quiet and looking the other way.[78] In this way,
Schwarzschlachten turned into a "collectively committed crime," although
that did not reduce the penalties for it. In some cases, entire families went
before the court; in some rural regions, the prisons were filled to overflowing
with the accused.[79] The number of suicides climbed. The rest saved themselves with gallows humor, joking that if one was not sent to the front,
then one would be behind bars for *Schwarzschlachtung*.[80]

Hitler himself was a vegetarian, and so if everything were to go according to his tastes, then soon nobody would be eating meat any longer any-
way. At supper with his entourage—as his secretaries later recalled—he
frequently described his visits to slaughterhouses in Ukraine in an effort to
spoil his guests' appetite for meat.[81] With the same fervor, he told his dining companions how farmers sowed seed, how the stalks grew, and how the
golden ears at last waved in the wind. Apparently Hitler believed that that
story would be enough to convert every committed carnivore immediately.[82]

Hitler had been eating vegetarian since the beginning of the 1930s.[83] He
did not eat any meat, not even broth, which according to Albert Speer
he called "corpse tea."[84] His meals could not be prepared with animal fat.
Later on, Christa Schroeder, one of his secretaries, reported that he har-
bored a downright "disgust toward meat" and was supposedly convinced
that "enjoying meat severs humans from a natural life."[85] Most likely, however, Hitler gave up meat not so much for ethical reasons but rather because
of his ideas about healthy nutrition. He believed that anything that was
cooked would make him ill and therefore swore by raw fruit and vegetables
and that, in any event, humans were supposedly vegetarians in vague pre-
historic times, and that was why they lived longer.[86] Hitler at times even
pondered whether he could get his German shepherds used to a vegetarian
diet—and, incidentally, the question of whether dogs can survive on a vegetarian diet is still being debated today.[87]

Nevertheless, despite his own convictions, Hitler was sure that he could not have meat-eating prohibited, because people would not understand such a prohibition. Only if more vegetables were provided and made affordable for everyone, he believed, would people then naturally make the switch to vegetarianism.[88] Indeed, Hitler could not stand it when someone gave up meat just to please him; he found it to be cheaply ingratiating, as Emmy Goering later recalled.[89] He did not miss the chance, however, for making his opinion on meat consumption freely known to others. Vegetarianism was one of his favorite topics. Time and again at supper he also told of Roman soldiers who purportedly fed themselves with plants and only ate meat in times of need.[90]

He was not alone in these views. Himmler, for example, intended in the long run "to limit the consumption of meat in future generations" of SS squads and to replace "meat and sausage products with something that satisfies the palate and physical needs just as tastily."[91] He gave very concrete nutritional tips as well: five times a week during the winter and at least three times a week at other times, people must eat "a warm evening meal in the form of soup, potatoes with skins on and then something extra that's chilled."[92] Bread was to be toasted in order to make it more digestible for those with intestinal ailments, but salted potatoes were to be "most strictly forbidden."[93]

A more or less ethically motivated vegetarianism first developed in the eighteenth century in Great Britain. Over the course of the nineteenth century, it became important in Germany too. It was in no way a purely *völkisch* ideal. In the early animal rights movement, there were also numerous left-wing groups that committed to vegetarianism and to renouncing experiments on animals, including, for instance, the Gesellschaft zur Förderung des Tierschutzes und verwandter Bestrebungen (Society for the Promotion of Animal Welfare and Related Endeavors). Founded in 1907 by the author and pacifist Magnus Schwantje, it was the first animal rights group in Imperial Germany. In 1918, the organization changed its name to Bund für radikale Ethik (Federation for Radical Ethics). Schwantje's credo was "Ehrfurcht vor dem Leben" ("reverence for life"), a sentiment that expressly applied to the lives of animals.[94] He viewed causing other beings as little suffering as possible to be the "supreme command of morality."[95] "Anyone who observes the behavior of animals thoroughly," he wrote in his book *Sittliche Gründe gegen das Fleischessen* (Moral Reasons against Eating Meat), "will recognize the grave injustice in allowing animals to be slaughtered just to provide themselves with lip-smacking enjoyment."[96] Schwantje called for limiting the breeding of cattle and pigs in particular: "Certainly no economic crisis would ensue if today thousands of Europeans immediately changed over to vegetarian food. On the contrary, every limitation on meat production will exert a favorable influence on the national economy."[97]

The Nazis liked to act as if they were the only true animal lovers; they did not closely examine the contradictions between their own animal welfare efforts and their radical actions against existing animal rights organizations. Most of these organizations were disbanded during the mid-1930s, including the Bund für radikale Ethik.[98] Schwantje had to abandon his work in 1933. In 1934, after he was arrested and interrogated by the Gestapo, he fled to Switzerland.[99] Shortly after the seizure of power, the Nazis also set about restructuring animal welfare throughout Nazi Germany. In April 1933, the Gesetz über das Schlachten von Tieren (Law Regarding the Slaughter of Animals) was proclaimed. It stated that now "during slaughter, warm-blooded animals are to be stunned before bloodletting."[100] This inclusion of all warm-blooded animals in the law meant that even chickens and other birds had to be stunned before slaughter. What at first glance appeared to function as a change in the law to the benefit of animals, however, had an ulterior motive, which was to prohibit Jews in Germany from practicing the ritual of shehitah, which features a specific method of slaughtering animals.

Nazi propaganda liked to represent Jews as being cruel to animals. An August 28, 1938, article in the Viennese broadside *Kleines Volksblatt* (Little national newspaper) reported, for instance, that shehitah showed the "overt, brutal, and unswerving cruelty of the Jew."[101] The Nazis in turn staged themselves as being especially kind to animals, propagating the notion of a "German *Volk*" who were culturally superior, who supposedly desired the safeguards of "humane slaughter."[102]

Criticism of shehitah was not new; it had come up during the Wilhelminian Empire in the course of intensifying antisemitism.[103] Thus, at the start of the 1930s, there were already prohibitions against ritual slaughter in twelve German states. Bavaria introduced such a prohibition in October 1930. Prussia and Hesse constituted exceptions, as a large proportion of the Jewish population lived in these two jurisdictions.[104] But the 1933 law that made stunning animals first a requirement changed all that. Granted, shehitah was not expressly forbidden—only slaughter without stunning—but at that time there was no method for stunning recognized by rabbis.[105] The legislation was a defining component of National Socialist *Judenpolitik* (policy on Jews) and one of the first incursions into the daily life of German Jews.[106]

How little the Nazis' animal welfare propaganda really meant to them was also brought to light during the war. On their campaigns to conquer Europe, the German Wehrmacht and the Waffen-SS also depended on foreign soldiers who were not necessarily always Christian. Tens of thousands of Muslim mercenaries, for instance, were fighting in the Balkans and in Eastern Europe, and the Nazi leadership made numerous concessions allowing them to practice their religion. In the rations they took care not to give them either pork or alcohol. From 1943 forward, they were moreover allowed to practice ritual slaughter without stunning, as were Muslim prisoners of war in Germany later on.[107] If it was a matter of retaining power and winning the war, then the Nazis' willingly abandoned their own laws.

Shehitah was not condemned on ethical grounds, nor did the Reich promote animal welfare for ethical reasons. What their animal welfare policies demonstrated above all else, as did the entire, diffuse Nazi *Weltanschauung*, was an effort to reorder the German *Lebensgemeinschaft* (community-based way of life) according to their own notions.

Against the backdrop of history, the regulations of the Reichsbahn (German Reich Railway) for transporting animals for slaughter are among the most drastic testimonies to this worldview. Large livestock and pigs were shipped separated by sex. If shipping were to take longer than thirty-six hours, the animals had to be given food and water beforehand.[108] The cars had to be well ventilated, so that the animals would not suffocate. How many pigs might be loaded into any car was precisely determined according to size and weight, so that the animals were able to lie down during the trip.[109] On top of that, there were appeals to breeders', handlers', and railroad workers' sense of duty: "Everyone involved in the shipment of animals has the duty of also seeing to it that the animals reach the destination station in the best condition."[110] Especially during the hot summer months, pigs could not be loaded into the cars if they had been overfed, since they would otherwise suffer a heart attack. A 1940 issue of the *Zeitschrift für Schweinezucht* showed what happened in that case, printing an image of about ten pigs that had perished in the train car during the trip.[111] Within just a couple of years, people would be jammed into these same freight cars and shipped to the east.

3

Drawing the Curtain
on Larval Stages

If you call a man a bug, it means that you propose to treat him as a bug.

—Aldous Huxley, *Eyeless in Gaza*

Victory seemed near. If the pupils in the classroom kept very quiet,
then they could even hear it. It sounded like clattering rain. The
sound came from next door. There, in a narrow closet, stood a
long set of shelves housing myriad white caterpillars tenaciously munching
through masses of green leaves. A good month ago, they had hatched out
of their tiny white eggs. In the beginning, they had been just a few small
millimeters in size and light as a feather. Since then, every last one of them
had molted four times and gained its weight tens of thousands of times
over. Now they were nine centimeters long and munching away stalwartly.
The crackling of their chewing apparatus was so loud that the pupils could
hear it through the closed doors.[1]

Though this particular scene did not take place, one very much like it
no doubt did across Germany. Let's have a closer look at one of the children in the classroom, listening to the caterpillars. Not an especially remarkable child but rather a very typical German boy. Let's assume his name was
Hans, because that was one of the most common boys' names in Germany
at this time.

Hans knew that it was of the utmost importance that the caterpillars
thrive. According to his teacher, raising them was supposedly a "kriegs-
wichtiger Dienst" ("essential service in the war effort"). That is why Hans
and his fellow pupils had to pick leaves from mulberry bushes in the school-
yard multiple times a day.[2] The little yard was not enough for the five
hundred trees that every school was supposed to plant, so mulberries also
lined both the athletic field and the market square located nearby, as well
as railroad lines and multiple boulevards.[3]

It looked this way in many municipalities in Nazi Germany, owing to a command issued by Bernhard Rust, the *Reichsminister für Wissenschaft, Erziehung und Volksbildung* (Reich minister for science, education, and national training). He saw the main task of the schools to be training National Socialists. One could become a model Nazi even at a young age, after all. For this reason, in June 1936, Rust decreed that as many schools as possible were to cultivate mulberries and train their teachers in sericulture. From the shores of the North Sea all the way to the Burgenland in the Ostmark (the "Eastern March," also known as Austria), more than twenty thousand schools would follow his call, plant mulberry seedlings, and take care of the little animals every day.

The caterpillars depended on the white mulberry; it was the only host plant for the silk moth. And these caterpillars could decide the war, making Germany's victory possible—at least that was the ministry's idea. For the insect larvae produce a material that no artificial fiber could beat—pure silk. It hardly ever catches fire and is water repellent; it is also extremely elastic and at the same time tear-proof. It was indispensable in producing parachutes for the German Wehrmacht.[4]

It was June 1941, about two years since Germany had been at war, and German troops had just invaded the Soviet Union. To Hans, who had never seen anyone die, the war likely seemed like just one big, faraway adventure. For the time being, while Germany's soldiers were fighting across Europe, there was nothing for him to do other than, like his classmates, to keep participating at home in the *Erzeugungsschlacht* (battle for agricultural production). Richard Walther Darré, the former *Reichsernährungsminister*, launched this program in 1934; food for the nation in his vision would come from its own soil, enabling Germany to free itself from dependency on imports from foreign countries.[5] And schoolchildren played a decisive role in realizing this vision, "helping with the achievement of the goals of the four-year plan," as the journal *Deutsche Wissenschaft, Erziehung und Volksbildung*, the official organ of the Reichswissenschaftsministerium (Reich Ministry for Science), stated.[6] In fact, as a school superintendent from Cologne wrote in 1940 in the periodical, "the economic blockade by our opponents in the present war demands the schools' intensified commitment."[7]

Silk farming had a long tradition in Germany. In the eighteenth century, King Frederick the Great of Prussia had promoted it in order to strengthen trade and commerce, admittedly without any lasting success. The Nazis suggested that this failure owed above all to the fact that people at that time were not passionate enough about the matter and that they had to an extent been coerced into silk farming.[8] Under the Nazis, sericulture experienced a boom like never before. The process was precisely regulated. The prices for eggs and cocoons were determined by the state, and silk farmers could only obtain their eggs from the Staatlichen Versuchs- und Forschungsanstalt für Seidenbau (State Experimental and Research Institute for Sericulture) in Celle, in Lower Saxony. They then had to deliver all their cocoons either to the Seidenwerk-Spinnhütte AG (Silkworks-Spinning Mill Corp.) located in Celle or to one of its collection sites, which could be found everywhere in the Reich. Those who contravened these guidelines and operated their own farm, for instance, had to pay up to ten thousand Reichsmarks in fines.[9] In particular, according to the program's provisions, people who wanted to earn a little something on the side by raising small animals or who were unable to carry out heavy physical labor were supposed to devote themselves to silk farming. Schoolchildren fell into both of these categories.[10]

April is when the first leaves sprout on mulberry trees. The first silkworms hatch from their eggs beginning in June, and for four weeks long they gorge themselves. Because the time for raising them can stretch into September under favorable weather conditions, schoolchildren could raise up to three generations of caterpillars in a row.[11] Still, caring for the caterpillars was anything but a child's game. Because the animals eat around the clock, the students would have to head out to get fresh leaves multiple times a day, which meant instruction time was lost. Late in the evening, the teachers would take care of the larvae one more time.

Hans was fascinated by the caterpillars, and, hungry for knowledge, he sought out information about them. He could be awakened in the middle of the night, and he would still immediately be able to recite what raising *Bombyx mori*—the scientific name for the silk moth—came down to, including that when it rained, it was necessary to dab the mulberry leaves off before they were fed to the caterpillars, because foliage that was too wet could give them intestinal catarrh; that between eighteen and twenty-three degrees Celsius was the best temperature for caterpillars, that is, neither too cool nor too warm but, above all, not too humid and muggy; and that it was imperative to keep their habitats clean so that no disease would spread.

For Hans and the other children, raising silkworms was a more or less exciting addition to their other lessons at school. It had a much more profound meaning, however, as the enormous political interest in the animals suggested. Participating in sericulture was not only meant to develop the pupils' virtues but was also supposed to convey to them Nazi racial doctrine and the "relentless character of nature."[12] The concept of racial purity assumed a central position in the Nazi curriculum and was brought home to children by way of plants and animals.[13] The pupils learned how to distinguish the individual caterpillar breeds and were instructed that they must never mix them up. They knew that they immediately had to cull caterpillars that turned lemon yellow, the ones that rooted around restlessly shortly before molting, or those that looked as if the air had literally gone out of them. The teacher explained that winnowing out all the sick and weak specimens was necessary to good cultivation.[14]

"School Materials"

Hans belonged to the generation that really and truly grew up during the Third Reich. In 1935, when he entered first grade just shy of seven years old, the German educational system had been *gleichgeschaltet* (co-opted by the Nazis) for some time. During the initial years of the regime in particular, the Reichswissenschaftsministerium did everything in its power to introduce the Nazi mindset in schools. Those teachers having a Jewish, Communist, pacifist, or otherwise critical stance were let go. And from the middle of the 1930s on, bringing pupils into line became the primary concern.

Hitler, who himself was in fact no model pupil, had clear expectations regarding this new way of schooling. In his opinion, it should provide "only general knowledge"; everything else supposedly confused children and overworked their brains. He thought learning multiple foreign languages was superfluous and that students should only take specialized courses in subjects they were interested in pursuing: "Do you see the necessity for teaching geometry, physics and chemistry to a young man who means to devote himself to music? Unless he has a special gift for these branches of study, what will he have left over of them later? I find it absolutely ridiculous, this mania for making young people swallow so many fragmentary notions that they can't assimilate."[15] For him, children's physical development was much more important: "What is weak must be pounded away," he said during one of his many table talks. "I will have them trained in all manner of physical exercise.

I want athletic youths. That is what comes first and foremost. That is how I will stamp out thousands of years of human domestication. In that way I will have before me Nature's pure and noble materials. In that way I can create what is new." Hitler wanted "a youth that the world will be frightened of."[16]

Since in Hitler's view a healthy body counted more than an alert, critical mind, up to five hours of sports every week was part of the class schedule.[17] In addition to these, there was a heavy emphasis on subjects whose purpose was "gesinnungsgebend" ("to provide conviction"), like German, history, and biology.[18]

Racial topics were part of the instruction for all school subjects, to be sure, and even found their way into mathematics, as the following arithmetic exercise from a 1941 math book for middle school demonstrates: "What probability exists for the appearance of a Jewish feature from a marriage of two mixed-race persons of the second degree?"[19] It was specifically the subject of biology, however, that was to impart systematically the "scientific truth" of Nazi ideology. As stated in the professional periodical *Der Biologe* (The biologist) in 1936, "National Socialist thought is necessarily biological thought."[20]

The task of the secondary schools as prescribed by the Reichswissenschaftsministerium in 1938 was to impart racial doctrine across all subjects in a "scientifically documented" manner so that pupils would come to learn that "all culture is racially conditioned."[21] The curriculum for history instruction, for example, listed the knowledge of the "superiority of the Nordic race," manifested from the deeds of Charlemagne and the conquests of the Vikings to the "*völkisch* significance" of Martin Luther, as a learning objective.[22]

In primary school, on the other hand, exerting a subliminal influence on the children was the method of choice for inculcating them.[23] To accomplish this, the Nazis set their sights on traditional teacher-centered instruction in particular. The teacher, especially the primary school teacher, occupied the role of a führer. He was not supposed to teach the children to think for themselves but rather to obey him unconditionally, uncritically, toeing the party line. "You are nothing, your *Volk* is everything!" was how the constantly repeated motto went. This emphasis on the group over the individual was evocative of how insect societies operated.[24]

And so sericulture seemed ideal for this form of pedagogical indoctrination: "How seldom do we in school reap the tangible harvest of industry, loyalty and conscientiousness, as swiftly as we see it presented here," wrote one school superintendent in a report on sericulture. "In all this work, many a child, who is otherwise never noticed, proves to be a useful member of the community."[25]

The otherwise unremarkable Hans and his classmates had a successful year raising silkworms. No diseases appeared during the summer; there were hardly any losses. Soon enough the caterpillars would seek out a place in the wooden latticework that the pupils had placed perpendicularly between the shelves. There, each larva would weave one single strand with its silk glands, located on its mandible, into a tight cocoon of countless, monotonous figure eights. It took them three days to complete their protective casing. If the life cycle of the insects were not interrupted, then they remained inside their cocoon for three weeks, during which time they turned into pupae before finally appearing as silk moths.

Not a single moth would emerge from these cocoons, though. A few days after the caterpillars spun their cocoons, Hans and his classmates detached them from the wooden frames and brought them to the central collection site. From there, they were sent to Seidenwerk-Spinnhütte boiled in huge cauldrons, and then skimmed from the surface in order to procure the silk.[26] It was only with intense heat that the silkworms released their precious commodity—one single silk strand, ten times thinner than a human hair and up to four kilometers long, of which up to nine hundred meters was usable.[27] To produce one single parachute took around fifteen thousand cocoons. In the end, all that remained were the burnt-up pupae destroyed inside them.[28]

Indoctrinating the youth did not end when school closed but also reached far into private life. During the first years of the regime, membership in Nazi youth organizations was still voluntary to a large extent. This changed, however, in March 1939. From that point forward, all boys and girls were legally "obligated to serve in the Hitler Youth" from age ten to eighteen.[29] Through this ordinance the Nazis would soon succeed in conscripting more than eight million of the approximately nine million youths in the country to the Staatsjugend (State Youth).[30]

Whereas the girls were to be prepared for their roles as mothers later on, for the boys it was all about training future fighters. At the Nazi Party's Nuremberg *Reichsparteitag* (Reich Party rally) in September 1935, Adolf

Hitler announced what he expected of them: "In our eyes the German boy of the future must be slim and slender, swift like greyhounds, tough like leather and hard like Krupp steel."[31]

Besides camping, field training exercises, and *Heimnachmittagen* (home club afternoons), Hans was especially fond of the *Jugendfilmstunden* (movie-time for youths). In these programs, the soldiers of tomorrow were shown propaganda films about successful battles.[32] Before the main film, *Deutsche Wochenschau* newsreels were run, which likewise would only report on the successful battles. Hans had recently seen his first heroes, the paratroopers, returning home, on one of these newsreels; they had captured the island of Crete from the British in the spring of 1941. In his mind, Hans could still hear the announcer Harry Giese. The voice of the *Wochenschau* raved about "unique military feats" and about "the toughest struggles against the strongest superior forces of the enemy."[33]

As Hans watched in astonishment as the soldiers marched through the rows of the cheering crowd, he may have just been thinking about how their parachutes had been woven from the silk of those caterpillars he had raised. A men's chorus resounded from the loudspeakers: "We only know one thing, when Germany's in need / Then to fight, win, and die must be our deeds."[34] For many boys in Hans's generation, death was at that time vague and mysterious, something that only affected others. They were raised in the belief that German soldiers were the best and the bravest and that ultimately they were invincible. Some boys even hoped that the war would continue indefinitely so that they might get the chance to prove themselves.

By this time only a very few would know what the purported *Helden-tod* (hero's death) really meant. Their most heroic accomplishment by this point in their short lives would presumably have been their passing the *Pimpfenprobe*, the test that every ten-year-old boy had to take before being admitted into the Deutsches Jungvolk (the German youth organization for children between ten and fourteen). It was made up of multiple parts that tested both athletic ability and rote memorization skills: they had to run 60 meters in less than twelve seconds, jump a length of 2.75 meters, throw a baseball twenty-five meters, recite by heart the "Horst-Wessel-Lied," the Nazi anthem, and "Vorwärts! Vorwärts!" ("Forward! Forward!"), the Hitler Youth's analogue to the "Star-Spangled Banner," as well as the *Schwertworte* (vows; literally, "sword words") of the Jungvolk, which went:

Young people are tough, tight-lipped, and true,

Young people are comrades,

Young people's supreme achievement is honor.[35]

For their final test, the aspiring *Pimpfe* (youngest male members of the Nazi youth organizations) had to undertake an outing lasting one and a half days.[36] This outing was very often the first time in their young lives that they had gone away from home on their own. They found themselves alone and subject to the whims of their *Jungenschaftsführer* (youth führers) who, though only a few years older, had already mastered the principle of commanding and punishing. Anyone who broke down during the march would be pushed that much more. Anyone who got homesick and let it show was considered a mama's boy and would be bullied even more.

Because Hans most definitely wanted to belong to the club, he would grin and bear it.[37] It was worth the effort. After passing all the tests, he finally got the hunting knife that all new *Pimpfe* were handed as a reward for being admitted to the Jungvolk. "Blut und Ehre" ("blood and honor") was engraved on the blade.[38]

Giese's jarring voice ripped Hans out of his reveries. The *Wochenschau* announcer was raving about the "selfless commitment" of the paratroopers, about their "difficult struggle" and "proud victory."[39] Not a word about death, though. That seemed to be only in their songs.

The *Wochenschau* kept quiet about the German massacre of the local civilian population on Crete and about the fallen paratroopers. Of the fourteen thousand men who set out in May 1941, only ten thousand returned. In truth, the case of Merkur (Operation Mercury), as the landing on Crete was called, was a suicide mission, and it incurred such heavy losses that there would be no further large-scale parachute operations undertaken by German paratroop divisions for the rest of the war.[40]

Combat the Cosmopolitan Cabal

Even if they themselves could not fight for Germany, many boys in Hans's generation had the feeling that they were at least contributing some small share by raising caterpillars. And they were able to do even more, too, for on the other side of the mulberry hedges, where the farmer's potato fields were, a genuine *Volksfeind* (enemy of the nation) laid in wait. Like the silk moth, the potato beetle is an insect, and like the moth's silkworms, the beetle and its larvae devour leaves. The beetles, however, are out for potato plants. Unlike the useful silkworms, the beetles were considered *Schädlinge* (pests), as Hans learned in school, for they endangered *Volksernährung*, the project of feeding the nation. Was it possible that Hans recognized that this distinction ultimately made no difference when it came to survival? For even if those caterpillars were spinning silk for the *Endsieg* (final victory) and these larvae were devouring the harvest, in the end both bugs would be exterminated.

Anything else Hans knew about potato beetles came from the skinny book that he got at school—*Die Kartoffelkäferfibel* (The potato beetle primer), it was called. By way of the rhymes in it, the children got to know where the beetle came from, what it looked like, and how best to combat it. They had to learn the entire book by heart. The teacher had each student recite lines in in succession. One of them started, then their neighbor had to continue the verse, and so forth. Hans's verse was

Everyone needs keep constant vigil,
and be prepared for defensive battle!
Only united can come success
in vanquishing this noxious pest.[41]

Hans learned from the primer how the beetle had made it to Germany in the first place. It was native to North America where it was first discovered in the Rocky Mountains at the beginning of the nineteenth century. Although until that point it had carved out a niche existence in the mountain valleys of Colorado, feeding on the leaves of the buffalobur nightshade plant, once the potato was being widely cultivated, the bug would expand its horizons. The North American beetle seemed to have been waiting just for this South American tuber. In 1859, it first stripped bare entire fields in the Midwest, around nine hundred kilometers east of its original range. Not even twenty years later, it reached the East Coast. With the help of the wind, entire swarms had been able to cross the Mississippi and Missouri rivers as well as the Great Lakes. On top of that, individual bugs had traveled even further as stowaways on trains and ships. It was only a question of time, then, before they would also cross the Atlantic and reach Europe.[42]

The agricultural ministry in Prussia was forewarned and had therefore preventatively prohibited potato imports from the United States in 1875.[43]

The very next year, however, the first Colorado potato beetles popped up in Bremen and other European harbor towns, and from there, they traveled on cargo barges unnoticed and ended up inland. Then, in 1877, just when zoologist Alfred Brehm declared "the fear of any introduction of this vermin to be baseless," the first reports of potato beetles in the Rhineland and in Saxony appeared.[44] After this initial incursion of beetles had been managed, though, German potato farmers would enjoy a good deal of peace and quiet for several decades.[45] Until 1914.

One of the many colorful drawings in Die Kartoffelkäferfibel shows little black-and-yellow striped beetles flying over a potato field in small propeller planes and shooting their guns at fleeing potatoes. Their squadron leader is wearing a bicorne, also called a "Napoleon hat," decorated with a tricolor feather of blue, white, and red. The message behind the bold drawing is easy to understand. Even primary school pupils would have recognized the real perpetrators of the plague immediately—not for nothing did old people mostly just call the beetle a "Franzosenkäfer" ("Frenchman's beetle").

The potato beetle had received its French appellation during World War I. At the time, the rumor arose in Germany that the French were supposedly propagating them on purpose in order to destroy German potato farming.[46] In point of fact, the French as well as the British contemplated deploying the beetles as a biological weapon but then rejected the idea out of concern for their own agriculture. The French, in turn, designated the Germans "doryphores" ("potato beetles") after they invaded in 1940 because they supposedly stole potatoes as well, so the story went. The motto "Combattre les doryphores" ("Fight the potato beetle") thus evolved into secret code among the Resistance for acts of sabotage against the German occupying forces.[47]

Yet even without military aid, the beetle had long been on its way to going global. By the middle of the 1930s, it had already crossed the western German border. Thereupon, the Reichsnährstand, in its capacity as overseer of agriculture and agricultural policy, established the Kartoffelkäfer Abwehrdienst (Potato Beetle Defense Service), which from then on would educate the population in the affected regions, distribute potato beetle primers to the schools, and direct search parties in the fields.[48] In 1941, this agency had 650 employees throughout the Reich who managed croplands of around ten thousand square kilometers, an area approximately ten times the size of Berlin.[49]

In addition, in 1940, an outpost of the Biologische Reichsanstalt für Land und Forstwirtschaft (Reich Institute for Agriculture and Forestry) was established in the Eifel Mountains, not far from Belgium. This so-called *Kartoffelkäfer-Forschungsstelle* (potato beetle research site) tested insecticide for combatting the beetle. While the approach to managing infested plants had previously been to cut them down, throw them into ditches, and then cover them with crude benzene and torch them, by this point the affected fields were being sprayed with calcium arsenate.[50]

However, since the beetle quickly became inured to the poisons, the surest way was still to gather them up by hand.[51] In some municipalities, every family had to put up one member for beetle-searching duty. Mostly these searchers were women and schoolchildren, whom you would see stooped over, stepping through the fields, eyes lowered.[52] Hans, too, would have been assigned searching detail once a week after school. Every pupil took on one furrow in the field, looking up and down the rows, to the left and to the right, plant by plant, leaf by leaf. The beetles could be easily recognized on account of their yellow-and-black coloring, and as a rule, they were usually found on the leaves and stems or in the leaf axils. What was much more laborious, though, was discovering their little yellow eggs, which were always stuck underneath the leaf.

The children were motivated in their search by the prospect of being awarded a potato beetle badge of honor for being the one who discovered the very first potato beetle in the field. All the other discoverers received a less decorative pin.[53] The beetles were collected in tin canisters and canning jars and subsequently annihilated. Grabbing hold of them was not difficult, but their black-and-red larvae squished easily between one's fingers, leaving behind a greasy yellow slime. If the children discovered eggs or larvae, then the representatives from the defense service came to spray poison.

The work of the children and other helpers met with success. After the potato beetle had expanded almost to the Weser River in central Germany during the 1930s, it was driven back into the southwest by the end of 1943.[54] It was a success that could also be used for propaganda purposes.

Creating an Enemy Image

Sericulture and books about potato beetles were not the half of it when it came to indoctrinating children like Hans. On their bookshelves were not only the potato beetle primers, adventure stories by Karl May, and the

Brothers Grimm's fairytales but also *Der Giftpilz* (*The Poison Mushroom*) and *Der Pudelmopsdackelpinscher* (The Poodlepugdachshundpinscher), both of which were written by Ernst Hiemer, the editor in chief of the anti-semitic weekly *Der Stürmer* (The Attacker), composed with the declared intention of conveying even to the youngest that there are "pests" among people, too.[55]

In *Der Giftpilz*, first published in a run of seventy thousand copies in 1938, colorful drawings portrayed Jews as being greedy, lice infested, inclined to pedophilia, and German hating.[56] *Der Pudelmopsdackelpinscher* is a collection of short stories about various animals with "Jewish" characteristics. The story of the wily, dirty mongrel dog of the title is accompanied by other stories featuring insects, such as bee drones, locusts, and caterpillars.

Anyone who as an adult can remember the bedtime stories of their childhood will be easily able to imagine that Hans's generation, too, would still have recalled these individual anecdotes for a long time after being read to. They would have remembered how in *Der Giftpilz*, a mother explains to her child while they are looking for mushrooms in the woods that just as poisonous mushrooms can pass as good mushrooms, so too can Jews, making it hard "to recognize Jews as crooks and criminals"; how the blood-sucking caterpillars in *Der Pudelmopsdackelpinscher* are compared to the Jews, who are described as representing "the same danger to humans as the caterpillars do"; and how because the bee drones make life so difficult for their people for so long that the other bees finally join together to kill them.[57] In the case of the drones, Hiemer is explicit in drawing the parallels: "There are not only drones among the bees; there are also drones among humans. They are the Jews!"[58] Even in the Grimms' fairytales there is a story called "Der Jude im Dorn" ("The Jew among thorns") that portrays the Jews as underhanded and rapacious.[59] The message behind these grotesque portrayals was equally as disgusting, namely, that the source of all evil lay in the Jew, the "eternal parasite."[60]

The term "parasite" had a long history, even before it was used as a label for people and, finally, as a pretense for exterminating them. In ancient Greece, the term still designated a highly regarded citizen; "parásitos" meant something like "tablemate" or "dining companion," indicating a state official who participated in the religious sacrificial feasts in the temple.[61] From that sense, a permanent and main stock character in Greek comedy subsequently developed, until "parasite" finally entered European languages during the sixteenth century, albeit with a negative connotation.[62] It was used disparagingly, as a designation for someone who lives at the expense of others. In the eighteenth century, the term made it into botany, where in English it was used to refer to certain plants.[63]

In the eighteenth century, Johann Gottfried Herder, the philosopher and member of Goethe's Weimar circle, was most likely the first to apply this newer meaning to people and to Jews in particular. In his work *Ideen zur Philosophie der Geschichte der Menschheit* (*Outlines of a Philosophy of the History of Mankind*), which Herder began publishing in 1784, he described them as "parasitic plants, having fixed themselves on almost all nations of Europe, and sucked more or less of their juices"; he did not consider the behavior to be any "jüdische Ureigenschaft" ("primal Jewish characteristic"), however, but rather the result of centuries-old ostracization of Jews.[64]

As is well known, it was the Nazis who brought the "scientification" of hatred for the Jews to a head. Countless examples can be cited, from book-length treatises to individual, hastily thrown together phrases, such as, for instance, when in *Mein Kampf* Adolf Hitler describes the Jews as a "bacillus" that "is expanding ever further," or when Heinrich Himmler speaks of the "corrupting plague in our national body."[65] The message of these comparisons was unambiguous; this danger had to be actively combatted. "With anti-Semitism it is just like it is with delousing," Himmler declared in front of SS corps leaders in April 1943. "Getting rid of lice is not a question of ideology. It is a matter of cleanliness. . . . We shall soon be deloused. We have only 20,000 lice left, and then the matter is finished within the whole of Germany."[66]

The Nazis' smear campaign against the Jews was not restricted just to bellicose writings or secret speeches either.[67] In addition to pedagogy, the Nazi regime also used pop culture to spread the image of the world they desired. Minister of Propaganda Joseph Goebbels relied heavily on films such as the 1940 *Jud Süss* (Süss the Jew) to sell the German *Volk* on the idea that Jews were the source of all evil. Also released in 1940, the propaganda film *Der ewige Jude* (The eternal Jew) shows hordes of rats streaming through streets and cellars while the announcer—Harry Giese, once again—narrates: "They are underhanded, cowardly, and cruel and occur mostly in huge droves. Among the animals, they represent the element of insidious, subterranean destruction. No different from the Jews among humans."[68]

Yet it was not only the Jews who were equated with pests.[69] In October 1939, shortly after the invasion of Poland, the Propaganda Ministry directed German newspapers that it was necessary "to make clear to even the last milkmaid in Germany that all things Polish are equivalent to all things sub-human." To be sure, the articles supposedly were to sound out this notion only subliminally and "always only as a kind of leitmotif." They were to do so by way of set phrases like "Polish economy" or "Polish depravity" and to repeat the message "until every German subconsciously regards every Pole as vermin, no matter if they're a farm worker or an intellectual."[70]

About the language of the Third Reich, the philologist Victor Klemperer notes that "words and phrases can be like tiny doses of arsenic: they are swallowed unnoticed, appear to have no effect, and then after a little time the toxic reaction sets in after all."[71] And for whom would this poison be more effective than in those in whom it is infused from an early age?

The Cult of Youth

In the fall of 1943, Hans, the fictional protagonist of this chapter, turned fifteen years old and graduated to the Hitler Youth. He was now able to wear the beige-colored shirt with the red-and-white swastika armband on his left sleeve and he even had his leather boots resoled with the swastika tread that had become available for purchase a few years before. The idea for the tread had come from a sidebar note in the magazine insert for Hitler Youth members, *HJ im Vormarsch* (Hitler Youth on the advance). The accompanying article asked, "How do I leave behind visible traces of my National Socialist beliefs?" Hans's shoeprints when he walked across soft, sandy ground made him proud. For, as the brief contribution to the magazine said, "They will know you by the tracks you leave behind."[72]

Hans might have also seen in the *Wochenschau* how the soldiers at the eastern front continued to trek further and further westward. The losses were enormous, and new soldiers were desperately needed. In the summer of 1943, an SS division called Hitlerjugend was even created. As the name indicated, it consisted in large part of former Hitler Youth members, many of them born in 1926 and thus just barely two years older than Hans. The idea for it came from *Reichsjugendführer* (Reich youth leader) Artur Axmann. His idea met with great enthusiasm from Hitler because he had in fact been hoping for especially fanatic fighters among the seventeen-year-olds.[73] Soon, even Hans would no longer be engaged only in the metaphorical battle to ensure successful agricultural production; soon enough, sixteen-year-olds would be literally fighting for the *Heimat* (homeland).

At the end of September 1944, in order to recruit additional soldiers, the Erlass des Führers zur Bildung des Deutschen Volkssturms (Führer's Decree on the Formation of the German Home Guard) was enacted. "As in the fall of 1939, we now again stand all alone over against the enemy front," Hitler claimed. "In the districts of the Greater Germanic Reich, all able-bodied men from the ages of sixteen to sixty are to form the German *Volkssturm*."[74] As far as the military was concerned, the recruits came under the command of SS-Reichsführer Heinrich Himmler.

In fact, the previous year, when Hans was celebrating his fifteenth birthday, Heinrich Himmler had delivered a secret speech to approximately ninety SS officers in Posen (now Poznań). In it, he spoke of the fact that the army not only could draft sixteen-year-olds but could also consider drafting "fifteen-year-olds." To anyone who had any remaining doubts at all about taking that step, he added that it was "better for fifteen-year-olds to die than for the nation to die."[75] The boys from Hans's generation came to know nothing of this speech. Many of them enlisted voluntarily in the fall of 1944. Very often, the hunger for adventure, a sense of duty, and years-long indoctrination mixed together, outweighing their otherwise careless levity.

Had most of the movie theaters not been destroyed by Allied bombs already, these boys would have likely been able to see themselves on the screen in the *Wochenschau*. In the newsreel, thousands of young war volunteers from across Germany stood lined up in the courtyard of the palace in Potsdam, where Axmann personally inspected the ranks. A few of them had just turned fifteen. Their heads rigidly turned to the left, they would stand there, motionless, as the numbers for war volunteers from the individual Reich districts were read out: "The East Prussia division," Axmann began—and here a voice exclaimed "Present"—"reports 9,482 war volunteers of the Hitler Youth. . . . The Cologne-Aachen division"—here a clearly softer voice responded with "Yes!"—"reports 9,715 war volunteers of the Hitler Youth . . . The Moselle Country division . . . reports 6,112 war volunteers of the Hitler Youth."[76]

Subsequently, Axmann stepped up to the microphone and announced: "Today I report to you, my führer, that from the 1928 cohort of Hitler youths, 70 percent have answered the call to enlist voluntarily for war.

These youths' genuine volunteer spirit for war will animate the fighting morale on the battlefield."[77] In March 1945, a visibly weakened, intensely aged Adolf Hitler received twenty Hitler youths in the garden of the New Reich Chancellery, where he decorated them with medals for their engagement on the front and for their "bravery in the face of the enemy." The youngest of them was barely older than twelve.[78]

While even boys from the 1935 cohort were conscripted in April 1945, Hans, like many others, was detached to build defense installations. At the edge of a field, his company piled up an earthen wall and laid out barbed wire. On the way there, they passed by a destroyed street lined with mulberry trees. In February 1945, even though silk farming never yielded enough silk to enable Germany to cease depending on imports, the Reichswissenschaftsministerium still issued a circular requiring the schools, "in consideration of the importance of sericulture for the war, to see to it that the use and care of the mulberry plantings be secured."[79] The tops of some the trees the company went by had been blasted away. Only burnt-out stumps remained of others. For the time being, the days of silk farming were over. Silk was not needed to make shrouds for corpses.

The era of the potato beetle, on the other hand, continued along seamlessly. Toward the end of the war, they began to be found in unexpected regions of the country. As in World War I, the British, French, and Germans accused one another of having dropped potato beetles on enemy territory. In October 1943, in fact, around fourteen thousand beetles raised specifically for this purpose descended on the town of Speyer in Germany's Palatinate region. Yet they were not dropped from Allied airplanes but rather from German ones. It was a test conducted by the military to see whether the invertebrates could survive a drop from a height of eight thousand meters. Not many were recovered, just fifty-seven in fact; yet all of those had survived the plunge.[80]

This biological weapon was never deployed, however. As the western front edged ever closer in 1944, the potato beetle research site was moved eastward from the Eifel Mountains to Mühlhausen, Thuringia, for security's sake.[81] Because the defense service had been severely neglected during the last two years of war, the beetle was able to expand across large swaths of the country. In 1944, it populated large portions of Bavaria. Like the Allies, it would soon cross the Elbe River and, after the war, rapidly advance all the way to the Oder.[82]

What would have become of Hans, had he actually ever lived? Perhaps at first sight of the enemy he would have thrown down his rifle, because he would have known that it was pointless. And how would he then have later looked back at his youth? Perhaps he would have recognized it as a great betrayal or else felt guilty. Perhaps, though, he would have perceived it as the best time of his life and therefore would not have wanted to hear it criticized. But perhaps the last year of the war would have also been his last.

Even though he himself did not exist, there were still hundreds of thousands of boys like him. Many from this generation were killed in the very last days of the war or stayed missing in action forever. How many there were can never be said with certainty, but their numbers definitely reach into the tens of thousands. Better for the youth to die than the nation, Heinrich Himmler had said. In the end they were only "materials," as Hitler called them, for the collapsing regime. A means to an end, of not much more value than silkworms.

4

Morituri

Katzenjammer, o Injurie!
Wir miauen zart im Stillen,
Nur die Menschen hör' ich oftmals
Grauenhaft durch die Straßen brüllen.
[Katzenjammer"—what an insult!
In the silence we meow,
People are the ones I hear most
in mean streets, making a row.]

—Joseph Victor von Scheffel, "Lieder des Katers Hiddigeigei"
[Songs of the tomcat called Hiddigeigei]

His name was Mucius, like the legendary savior of Rome. Because a tomcat does not understand that much about classical heroism, though, they mostly called him Muschel. Not with the sibilant "sh" sound, the one when you want to shoo a cat off the oven but, rather, with a soft "j," like in the French word "jamais"—Mujel.[1]

The cat had been living with Eva and Victor Klemperer for more than eleven years. They had been married for thirty-six, never had had any children; they had never wanted any, either—not Victor, at least. He was bothered when Eva fussed over Muschel too maternally, as if it had been their child. Still, he was attached to the gray tomcat. Especially because the cat meant so much to Eva. It often seemed to him that only Muschel was able to cheer her up when she would withdraw for days at a time, plagued by depression, when Victor did not know how to get through to her.[2] In the spring of 1942, life in Dresden hardly gave any cause for hope that things could get better again one day. "We're totally isolated," Victor wrote in a letter to his sister. "Our only, most faithful contact is Mucius, AKA Mujel."[3] The tomcat lifted their spirits, one of the last certainties in an uncertain time in which the intolerable had been commonplace for a long while.

Eight years prior, in 1934, Victor Klemperer had been barred from the local cat club because he was not an "Arier" (Aryan). In 1935, for the same reason, he lost the teaching position he had held in Romance languages since 1920, at what was then Dresden's College of Technology (now its Technische Universität). Still, he and Eva stayed in Germany. Only after that night in November 1938, when neighbors became arsonists, did they decide to emigrate. They wanted, like so many, to go to America. Victor's brother, who had been living in the United States for a long time, vouched for them so that they could enter the country. The waiting list was long, though. Neither the U.S. consulate in Berlin nor the Jewish Community in Dresden was able to provide further assistance. So they applied to emigrate to other countries—to South Africa and Rhodesia, to Australia as well as Peru—yet they applied everywhere in vain.[4]

That is why they stayed. Until nothing was possible anymore and all the borders were closed. And, strange as it may sound, Victor Klemperer was somehow even relieved about it. Because at bottom he never wanted to leave—why go even? He was a German, after all. It was intellect that mattered to him, not blood.[5] Besides, they would have had to sell their house in the suburb of Dölzschen far below its value. Then, too, what would have become of Muschel? They were apparently willing to give him to a close acquaintance, but she had declined on the grounds that he would not get used to the change. To put him to sleep, she said, would be more humane. They could not do that. So they stayed. Until it was too late to go away.[6]

Then, in May 1940, they had to leave their home and move into one of the numerous *Judenhäuser* (Jews' houses) that were located all over town, and where they were entitled to only two rooms. The kitchen and the balcony off the bedroom they shared with the neighbors. They had been living there ever since, in constant fear of house searches by the Gestapo. When every so often friends gave them a fish head for Muschel, they boiled it up immediately and then burned the bones up so that the Gestapo would not get suspicious—eating fish in Jewish households had been forbidden for a long time now.[7]

Even outdoors Victor Klemperer was hardly able to move around freely. He was no longer allowed to enter the Baroque-era Großer Garten (Great Garden) in the city center; after nine o'clock at night he was not allowed to go out into the street. New prohibitions came in constant succession. In his diary, he noted, "Any animal is more free and has more protection from the law."[8] He even had to go to jail one week in July 1941, because one evening he had forgotten to darken one of the windows in his study. The most drastic regulation, though, was that starting on September 19, 1941, he had to attach a yellow star with the word "Jude" ("Jew") written on it to the breast of his coat. It seemed to him as if he were carrying his own personal ghetto around with him, "like a snail its house."[9]

All these humiliations made him increasingly bitter. As he had noted in August 1937, "There is very little feeling for people left in me. Eva—and then comes Mujel, the tomcat."[10] He had long become the symbol of their endurance. "The tomcat's raised tail is our flag," the two pledged to each other time and again: "We shall not strike it, we'll keep our heads above water, we'll pull the animal through." Once everything was over, Muschel was going to get a veal cutlet from Kamm's, the best butcher in the area, as a victory feast. At least that is what they had planned.[11]

Sometime in 1940, the Reichsernährungsministerium came up with the idea of forbidding house pets completely as a way to save food for people. Hitler did not think much of the idea, and he intervened immediately when he found about it. From his point of view, such a prohibition could not be expected to be accepted by the "animal-loving" German *Volk*.[12] For this reason, the Reichsernährungsministerium decided on a more nuanced solution, prohibiting only the keeping of animals in those households that did not count among the *deutsche Volksgemeinschaft* (the community of the German *Volk*). In the view of Nazi bureaucrats, animals in non-Aryan homes cost the nation foodstuffs and, on top of that, created a wrinkle in their deportation plans. The house pet prohibition, which applied in particular to Jewish pet owners, thus had an additional—logistical as well as cynical—ulterior motive: by prohibiting house pets, they would no longer have to worry about the house pets left behind when their owners were deported.[13] Transports to the east had been taking place since October 1941. At the beginning of 1942, there were still about 130,000 Jews living in Germany. One year later, there were only 50,000, representing one-tenth of the original Jewish population.[14]

It was from a neighbor woman that Victor Klemperer found out they would have to give up Muschel. Then it appeared in the newsletter for the *Jüdische Gemeinde* (Jewish community), too, in black and white: as of this time, Jews were no longer allowed to keep any house pets.[15] By this point in time, the Nazis had taken just about everything from Victor—his occupation, his house, his reputation, and his everyday routine. The death sentence for the tomcat was the final blow. That same evening, he wrote in his diary, "What a low-down, slimy act of cruelty towards a couple of Jews."[16]

Prohibiting house pets was a further step toward the complete disenfranchisement of Jews in Germany. After the directives that had driven them systematically out of public life and had forbidden them to leave the country, it was the next cut into an everyday life of which there would soon be little left over. As it stood, as a result of expropriations and forced moves into "Judenhäuser," withdrawing into private life was hardly possible anymore. The house pet prohibition was a deliberate attack on one of the last remaining sites of refuge. As a woman eyewitness living in Berlin would say later about these years, "House pets weren't able to play any political role, but for many they were the only living being that still greeted them happily at home after a day of forced labor."[17] Once the house pets were gone and their familiar sounds—whether it was the padding of their paws on the wooden floor or the scratching of their claws on the leather sofa—had faded from memory, then home lost its meaning.

Some simply refused to give up their pets, such as the family of twenty-year-old Herta Höxter from Nuremberg. Two years into the prohibition, they still had their two cats and were always at risk of being caught. In the summer of 1944, when the Gestapo came to the door, Herta had just enough time to shove both animals into the clothes closet. Fortunately, they did not make a sound, so the Gestapo did not notice them.[18]

In the occupied areas in the east in these years, people everywhere were being deported, including from the Romanian town of Carei on the border with Hungary. When, in 1944, the Jews from Carei were taken to Auschwitz-Birkenau, the local paper there appealed to the townspeople to take in the dogs and cats that had been left behind in the ghetto.[19]

Objects of Hate and *Herrentiere*

Even if this episode makes the Nazis' misanthropy especially obvious, their "decent attitude toward animals," as Heinrich Himmler once put it, was still firmly part of their crude self-image.[20] The Nazis boasted about their love for animals and their laws for protecting nature, which had garnered them international praise in the mid-1930s. Unlike in France and Great Britain, where laws primarily focused on the rights of domestic and working animals, in Germany, the new *Reichsgesetze* (Reich laws) also mandated protection for wild animals.[21] However, all animals were not equal under the law, as the case of cats makes especially clear.

As of March 1936, a special ordinance on the protection of nature allowed every property owner to capture stray cats on their premises and then turn them over to the local police. If the cat owner did not come forward within three days, the police could have the cat killed. If a cat was caught more than twice in one year in a stranger's garden, then it was "to be rendered harmless," as it was euphemistically described.[22] In addition, since 1934, hunters had had the right to shoot any cat dead that was located more than "200 meters from the nearest inhabited house."[23]

One of the most eager cat hunters was Will Vesper. The writer and journalist was additionally considered to be "one of the worst nationalist fools," as Thomas Mann wrote in a letter to his friend and fellow author Hermann Hesse.[24] Vesper was well known for composing animal fables and poems for the führer as well as for stirring up hatred for purportedly Jewish authors and publishers like Hesse in the Nazi journal *Die Neue Literatur* (The new literature). On walks with his dog through the expansive park of his estate, near Gifhorn in today's Lower Saxony, he shot dead all the cats he could find not only because he wanted to stop them from raiding the nests of songbirds but also because in his view cats were an alien, unpredictable race from the Orient.[25] In contrast to dogs, as he drummed into his little son Bernward, cats supposedly could not be instilled with any "human" character. They were, he thought, purely urban animals—deceitful, false, and asocial; in short, "the Jews among animals."[26]

Cats show how arbitrarily the purportedly systematic Nazi ideology was construed by its proponents, for even among the regime's adherents there were numerous cat lovers. Many of them tried to improve the negative image of the house cat by positively interpreting its desire for freedom. Its supposed untameability was proof of the fact it was a *Herrentier*, a master animal, since it would not submit to anybody.[27] In 1931, Ferdinand Hueppe—a *Rassenhygieniker* (racial hygienist, that is, eugenicist) and the first president of the Deutscher FußballBund (German Football Association)—had praised cats for being "our hygienic helpers in the national recovery," on account of the fact that they hunt mice and rats. At the same time, he complained that "in no other cultured country were [they] so basely mistreated and persecuted as in Germany."[28]

Friedrich Schwangart agreed. The zoologist and animal psychologist considered hatred of cats to be "a characteristic for broad strata of Germans," as he wrote in his 1937 book *Vom Recht der Katze* (On the rights of cats).[29] Schwangart was a cat expert who at the end of the 1920s drew up the first standardized breed descriptions for long-haired cats, earning him the nickname "Katzen-Schwangart" (roughly, "Schwangart the Catman"). Because he was an antifascist, however, he had been compelled to give up his honorary professorship at Dresden's College of Technology in 1933.[30] Even so, he would continue to criticize how the Nazis dealt with cats. He perceived crass contradictions between their treatment of cats and their "otherwise highly developed animal welfare policy and otherwise humane attitude toward animals." While they put mountain goats, wisent, and beavers under protection and gave their all to resurrect centuries-extinct aurochs, the domestic cat was designated as an exception to this general conservation rule.[31] In his writing, Schwangart attempted to speak out against these prevailing prejudices. In no way, he wrote, was the cat a "standoffish loner"; indeed, to the contrary, "the cat in particular is the house pet of the poor, of the materially and emotionally afflicted, and, often enough, their last happiness."[32]

When this last happiness was endangered, an already dismal existence was made that much worse, as in the case of the Klemperers. At even the thought of having to put their tomcat, Muschel, to sleep, Victor Klemperer would become violently ill. But he was much more worried about his wife's reaction to this loss. "It [i.e., Muschel] had always given Eva support and consolation. Now she will have even less resilience than before."[33] Now, of all times, when she needed it that much more.

"You still have your work," she reproached him, full of rage and sorrow, on hearing the news about the prohibition. "Everything has been taken from me."[34] He could not even think of blaming her for this rebuke. It was on his account that she had given up her calling as a concert pianist. Instead of composing music, she typed up his articles and manuscripts and corrected them. What Muschel was for her, Eva was even more so for him—he had only her and her alone to thank for the fact that he was even there at all, instead of deported a long time ago like so many of their friends, acquaintances, and neighbors. She was an insurance policy for survival; there was no "J" for Jew sticking out on her *Kennkarte*, the official domestic identification for the police. Instead, hers showed the Reich eagle along with the swastika in its oak leaf wreath. For Eva was an "Arierin" (Aryan woman).[35]

An animal that lived with one Jew, though, was still a Jewish animal, the way the Nazis saw it.[36] Because the Klemperers did not dare give their cat to somebody else and yet did not want to surrender him to the police either, the following day they decided to bring him to a veterinarian.[37]

A Life on the Margins

Hardly any other kind of animal is comparable to the domesticated cat. It had taken a long time for it to be accepted in Germany. Then again, as the Scottish writer Compton Mackenzie once opined, it was puzzling "why it ever decided to become a domestic animal" in the first place.[38]

The cat found a home in Europe much later than the dog. All of today's house cats originally descended from *Felis silvestris*, a subspecies of the wildcat, which had once lived in North Africa and Arabia. Its domestication began eight thousand years ago in Asia Minor. By the ninth century, the first cats had made it to Europe. During the High Middle Ages, knights returning from the Crusades presumably also contributed to their multiplication on the continent.[39]

Though they were tolerated as hunters of mice and rats both in the countryside and in the towns, for a long time they still had a bad reputation in Central Europe. Because they mated in full public view and the females also wailed loudly in the process, the cat was considered to be a symbol for witchcraft and suspected of being in league with the devil. Many also viewed the cat as an agent of the Black Death, that pestilent pandemic that raged in Europe during the middle of the fourteenth century. Christian clerics like the predicant Berthold von Regensburg maintained that cats bore the plague on their "poisonous breath," and a correlation was made between cats and Jews, who were believed to have made a pact with the devil and to have poisoned the wells. As a consequence of these epidemics, the Jewish population was subjected to pogroms in many places in Western Europe, and cats became part of the targeted hunts as well.[40] Killing cats turned out to be a mistake that had disastrous consequences: where the cats disappeared, the rats were able to expand unchecked and with them the rat flea, too, which was a carrier for the plague bacterium *Yersinia pestis*. In the end, this mistaken belief, brought into the world by the Catholic

Church, contributed to the fact that twenty-five million people fell victim to the Black Death.[41]

Over the course of the fourteenth century, the house cat became so rare that its presence was noticed and specifically mentioned in historical chronicles.[42] Its reputation hardly improved after that. Even the aspiring bourgeoisie of the eighteenth and nineteenth centuries, which regarded house pets as evidence of being cultured and civilized, shunned the cat.[43] In contrast to songbirds—which were said to have bourgeois characteristics, like being monogamous, caring, and musically talented—the cat embodied everything that was unbridled and immoral.[44]

Indeed, bird lovers were their fiercest opponents, traditionally speaking. This animosity continued into the Wilhelminian era. Hatred for anything foreign was readily associated with purported love for animals, and several bird conservationists were of the view that the cat was "never a genuine, German house pet" but rather an "immigrant enemy from the East."[45]

Cats did have one influential proponent, however. Alfred Brehm, who had imputed malice to many an animal, writes about cats in his *Thierleben* (Animal lives): "The higher a people, the more definitively they have become settled, the more widespread is the cat." On top of that, he contradicted the image of the cat as being insidious and hard to train. Under human care, even wild forms like the lion would become "often wholeheartedly tame."[46]

Because house cats never quite gave up their wild ways, various German towns at various times made an effort to tame them by means of legislation. So in 1911, for example, a municipal committee in Munich came up with the suggestion to levy a cat tax in addition to the one on dogs. Admittedly, it never got any further than just the idea, for the council members considered controlling the mostly wild animals to be simply impossible. Others, in turn, objected that a cat with a tag around its neck could get caught while climbing and so be easily injured. Several did not take the suggestion at all seriously and joked that in that case, "even Monday morning hangovers" from tomcatting around all weekend would "also [have to] be taxed."[47]

In 1930, the city of Dresden did make an attempt at it and introduced a cat tax, but it abolished the tax in the following year. As the *Dresdner Nachrichten* newspaper reported in March 1931, the revenues only amounted to a little more than a hundred thousand Reichsmarks—which would translate today to thirty thousand euros—which is to say, "downright pitiful."[48]

Cats would remain tax free and continue to conduct their lives on the margins of society.

Meanwhile, at the Klemperer residence, three days had passed since they received the news. Muschel was still with them. They had not had the heart to take him to the vet. Eva barely succeeded in getting out of bed in the morning. Muschel, on the other hand, was behaving more exuberantly than ever before. As she watched the tomcat, Eva said, more to herself than to her husband, "The little creature plays around, is cheerful, and doesn't know it's going to die tomorrow." Victor gave no response, though one thought immediately leapt to mind: if there is anybody who knows whether they will die tomorrow.[49]

Time was closing in on them; a written order to hand Muschel over was supposedly already on the way. Once it was delivered, what happened to him would no longer be in their hands. On the other hand, if the Gestapo were to find out that a tomcat from a Jewish household had been brought to the veterinarian instead of handed over to the authorities, it could mean the end for her husband.[50]

For his farewell, Eva got Muschel 450 grams of veal. That is nearly as much meat as Victor and Eva were allotted per week. Ever since the war began, foodstuffs had only been available with food tickets, and meat was strictly rationed. In the middle of 1942, the weekly rations for two individuals amounted to six hundred grams; soon enough Jews would no longer be allocated any meat at all. Muschel knew nothing of that, of course. He devoured the veal with relish, smacking his lips. It was his proverbial last meal.[51]

Then the fourth day, too, gradually drew to a close. Victor left it up to Eva whether she would take Muschel away or not. At four o'clock in the afternoon, an hour before the veterinary closed, she pulled herself together, put Muschel into a cardboard box, and set off with him. She petted and calmed him when the veterinarian injected the hydrocyanic acid from a syringe; she stayed with him until he no longer moved. When she returned home that evening, she only said these four words to Victor: "He did not suffer."[52]

Half a week later, they would find out from their neighbor Frau Kreidl that her Jewish husband had perished in the Buchenwald concentration camp. "Muschel died three days too early," Eva said to Victor. "Today he could officially belong to the Aryan widow Elsa Kreidl." Victor was hurt that Eva was still thinking about Muschel when he was fearing for his own fate.[53]

"Moriturus days"—the days of the one doomed to die—is what Victor Klemperer, in retrospect, called that time when they wrestled with whether they should hand Muschel over or not. On New Year's Eve, he would write in his diary that 1942 had been the worst year of all so far. Muschel was no more, and all those they had been together with a year ago had disappeared. It had been Moriturus days without end.

The Pull of the Predator

Regard for house pets in general changed fundamentally during the Third Reich, so even though cats were viewed with suspicion on account of their fickle nature, they were not singled out. In their critiques of civilization, psychologist Erich Rudolf Jaensch and zoologist and medical practitioner Konrad Lorenz, for instance, argued that these animals were not proof of how highly developed a society was but rather of how *entartet* (degenerate) it had become. What was wild and primordial was now at a premium, and that did not just apply to animals: "Since its beginnings, the Nordic movement has been directed against people being 'turned into house pets,'" Lorenz wrote in 1940. "It fights for any development that is, in fact, opposed to the direction in which today's civilized, metropolitan humanity is headed." Creating a healthy *Volk*, Lorenz insisted, supposedly required "an even more rigorous weeding out of the ethically inferior" than what was already happening at the time.[54]

Just as the city dweller was considered to be effete and reprobate in contrast to the peasant, so, too, the domesticated animal—with the exception of the "faithful" dog—was seen in Nazi ideology mostly as a degenerated form of the wild animal.[55] Even so, this notion did not make a dent in the popularity of house pets among the larger population. The men in the leadership ranks surrounding Hitler, too, kept house pets—primarily dogs, like their führer, a dog handler himself. In fact, not one of them had anything to do with cats. Apart from Hermann Goering.

Mucki (literally, "small bug") was his name, because in the beginning he was "as little as he was comical."[56] Since he would one day become a stately king, Goering also called him Caesar. That sounded more in keeping with his status, more majestic. Together with his wife, Emmy, Goering raised him from the bottle. At that time, Mucki was already as large as a house cat, in fact. From the size of his paws, though, it was apparent he was not done growing. Soon Mucki was approximately the size of a German shepherd.

The little brown circles on his sandy-colored coat, so typical for young animals of his species, were fading. For good or ill, before too long Goering would have to part with him, as he had also done with Mucki's predecessors. For a fully grown lion was a little too large and dangerous, even by Goering's housing standards.

The fact that he kept wild animals at home, just like Roman emperors had, accorded with Goering's whimsical nature.[57] He loved anything majestic, extravagant, and ostentatious. He collected titles, trophies, and paintings the way others collected stamps. Over the course of time, he served in more than twenty offices. Besides *Reichsjägermeister* (Reich hunting master), *Reichsluftfahrtminister* (Reich aviation minister), *Vorsitzender* of the Ministerrats für Reichsverteidigung (president of the Reich Ministerial Council of Defense), and *Oberbefehlshaber der Luftwaffe* (supreme commander of the air force), Goering also gladly assumed the position of *Reichskriegsminister* (Reich war minister). Yet after he successfully pushed the incumbent Werner von Blomberg out of that office, he went away empty handed, inheriting only Blomberg's mocking nickname—"Gummilöwe" ("rubber lion").[58]

As far as animals were concerned, Goering had clear predilections. He hardly knew what to do with dogs, though he had enough insight to understand they were indispensable for hunting. His fear of snakes sent him into a downright panic; on the other hand, he liked horses very much, even if he preferred to have them drawing a carriage rather than riding them himself. He loved shooting at deer and wild boar, though no creature admittedly affected him as much as the king of the beasts. For him, there were in essence only two zoological classes: "animals—and lions!"[59]

In public, too, Goering liked to present himself as a "fanatical animal lover."[60] When Mucki was still little, he had himself photographed by Heinrich Hoffmann—Hitler's personal photographer, who shot pictures of him at both work and play—as he was holding the baby lion in his arms, giving it a bottle. And when he was recovering at the mountainside retreat at Obersalzberg, outside Berchtesgaden, from injuries sustained in an automobile accident, the *Deutsche Wochenschau* cameras filmed him lying on his divan, reading the newspaper on the terrace, while Mucki crawled around him on the terrace wall. Neither Goering nor the cameras took much notice of the hunting dog that sat there quietly in Goering's shadow.[61]

Goering got his first lion in July 1933, as a present from the zoo in Leipzig.[62] Through 1940, a total of seven lions lived with the Goerings.[63]

The young animals each stayed a little more than a year in their household. When they got too big, he handed them off to his hunting buddy Lutz Heck, the director of the Berlin Zoo, who promptly supplied him with young reinforcements.[64] The numerous guests of state who came and went at the home of the number-two man in the Third Reich also liked to get their photos taken with the little lions—including the king of Siam, American aviation pioneer Charles Lindbergh, and Italian dictator Benito Mussolini. The era of big cats would not end in the Goering home until their daughter Edda, turning two years old, began to walk around on her own and it became too dangerous for her.[65]

Mucki always stayed close to Goering, no matter if he was at meetings at the Prussian State Ministry in Berlin, at his summer home on the Obersalzberg slopes, or at his hunting lodge called Carinhall on the Schorfheide in Brandenburg, northeast of Berlin. Both at Carinhall and at his official residence on Berlin's Leipziger Platz, Goering had a lions' den built, which his employees had to keep meticulously clean. In Carinhall, furthermore, the big cat had his own outdoor enclosure, though most of the time he ran around loose, following Goering's every step.[66] Even when he traveled to the Rominter Heath in East Prussia to go hunting, he took his lion along, as in January 1937.

The day they arrived, everything had been prepared. *Oberforstmeister* (chief forest officer) Walter Frevert, along with ten forestry officers and six policemen, had come on this cold, wet winter day to duly receive Germany's supreme huntsman in the inner courtyard of the *Reichsjägerhof* (Reich hunting estate). When the limousine came through the courtyard entrance, they quickly readied themselves. The local police made an effort to present their carbine rifles, even though their bellies got in the way a bit. The hunters had already put the bugles to their lips when the car came to a sudden stop. Goering's adjutant Karl Bodenschatz jumped out, waving his arms wildly. "Don't blow," he shouted at the men, "the lion will go crazy!" Shortly afterward, Goering hauled himself out of the back seat of the car, carrying Mucki in his arms. "Where can we pen him up?" Goering asked his top forester. "The bathroom would be best," Frevert said. Once Mucki had been stowed in the bathtub, they headed off to hunt wild boar. For hours, while Goering indulged his passion in the slush, the lion was peeing all over the bathroom.[67]

The Nazis venerated predators among animals most of all because they identified with them.[68] In their writings, thinkers like Friedrich Nietzsche and the historian Oswald Spengler involuntarily had provided the template for that identity with their metaphor of human beings as predators: "The animal of prey is the highest form of mobile life," Spengler notes in his work *Man and Technics* (published originally as *Der Mensch und die Technik* [1931]). "It imparts a high dignity to Man, as a type, that he is a beast of prey."[69] For Hitler, too, the predator is the leitmotif par excellence, as is evident in his vision of what future German youth would be like: "What is weak must be pounded away. I want my youth to be strong and beautiful, masterful and intrepid. The free, majestic predator must flash from their eyes."[70]

Though the house cat had a tough time of it, its larger wild relatives embodied everything that the Nazis treasured in predators. The lion was a sign of power and prestige, while other big cats were admired for their agility, aggressiveness, and speed and would end up playing a far more significant role than the lion. For if the future generation of soldiers was to be aggressive and agile, their weapons ought to be as well.

Ironclad Cats

At the end of August 1942, four "tigers" were trying to get through the swamps around Leningrad. These tanks hardly made any headway, however, as they were much too heavy for the boggy ground, and so they kept sinking into it over and over again. Strictly speaking, they were not at all ready for this operation, because they had still exhibited too many defects. Hitler had become increasingly impatient, however, and wanted to see rapid results. That was why during the armaments conference at the beginning March, he personally pushed for trying the tanks out right away on the front lines.[71] They were supposed to surprise the enemy and beat them resoundingly, but instead, their first operation ended in a disaster: three of them gave up the ghost in the middle of the bog and had to be laboriously pulled out with tractor rigs. The only thing worse than the failure of the panzer tanks was that the Russians would now know about the latest German secret weapon. And, above all, its weaknesses.

After the Russian campaign began in the summer of 1941, it became apparent that German tanks were not as effective as Nazi propaganda had tried to make people believe. Moreover, the Wehrmacht had not been configured for a long, persistent battle during the Russian winter, especially not at the minus forty degrees Celsius temperatures that marked the winter of 1941–42. During this time, almost one million soldiers lost their lives, from either freezing to death or succumbing to disease. Furthermore, by that time the Red Army had a tank at its disposal, the T34, that was superior to all German models, which terrified Germany's Wehrmacht. In November 1941, for the first time, more German tanks were destroyed than Russian ones. New and above all quicker panzers were needed that could compete with the T34.[72] And not just that—their very names had to make clear that they were as fast, nimble, and deadly as predators. That is how the Rüstungsministerium (Armaments Ministry) got the idea to name both new models after big cats. For the first time in the history of the German military, warcraft would not be designated with combinations of letters and numerals. The heavier of the two tanks received the name "Tiger" and the lighter, more nimble one was called "Panther."

One year after the first four Tigers got stuck in the swamp near Leningrad, it did not go much better for the Panthers near Kursk.[73] Victories became increasingly rare, until the war finally reached Germany, too. At first, it came from above.

During the course of the war, Dresden had already withstood many airstrikes. Victor Klemperer had lived through them all and survived, just as he had eluded deportation to that point. The "Moriturus days," as he called them, turned into entire years. The number of people in his sphere shrank. And with each additional day it became more dangerous for him as well. He knew about the rumors that even the last Jews remaining in town were to be "evacuated," as the Nazis officially called it. Klemperer had also heard about an external work detail, and he had an inkling of what that meant.[74] Though he did not know much about Auschwitz, he was aware of what would await him there. As of February 13, 1945, there were only seventy Jews still living in Dresden. They were supposed to be deported in the following weeks, and he would be among them then. Yet in the end it did not come to that.

It was around ten o'clock in the evening when he heard a faraway buzzing that droned louder and louder.[75] It was coming from the engines of British bombers on their approach to Dresden. Since the fall of 1944, there had been 174 airstrikes.[76] Still, the people of Dresden had not ever lived through what descended on the city this February night and for the next day and a half. The British and American squadrons dropped almost four thousand tons of explosive and incendiary bombs.[77] It set off a firestorm that subsequently raged through the streets of the city center, making no distinction between Aryans and Jews. And yet, as appalling as it might sound, it would be Victor Klemperer's salvation. In the chaos of the rubble and the fires, he and Eva were able to escape from the Gestapo.[78]

In the days that followed, they slogged their way south, via Pirna, into the Vogtland region and from there all the way to Bavaria, where they would learn about the end of the war three months later. In June 1945, they returned to Dresden. Soon after they moved back into their old house, which they had had to vacate five years before. It would not be long before it also became the territory of a white tomcat with gray spots.[79]

Two years after the war ended, Klemperer published his work on what he called the "lingua tertii imperii," which he had written secretly during the war, after he had been barred from the libraries and archives and could only work at home. The book was published under the title *LTI: Notizbuch eines Philologen* (*Language of the Third Reich: LTI—Lingua Tertii Imperii*), its title alluding to the Nazi predilection for abbreviations. In it he mentions that decree from the spring of 1942 that had meant the end for their tomcat Muschel as well as many other animals. How many pets from Jewish households this order cost is not known, but it happened, Klemperer notes, "not in individual cases or as isolated malice but, rather, officially and systematically, and that is one act of the cruelty that no Nuremberg trial reports, and for which, if it were up to me, I would erect a towering gallows, even if it cost me eternal bliss."[80]

At that time, of course, the gallows were far from being taken down yet; many perpetrators got away, nevertheless, and even got on with their (new) agenda. But the traces of the Third Reich were still perceptible; Klemperer found them everywhere in the German language. He noted how much the cadence of the speeches of Communist politicians in the GDR were similar to the National Socialists they hated so much. This "language of the fourth Reich," as Klemperer calls it, appeared to him to be "sometimes less different from that of the Third Reich than the difference between Saxon dialects spoken by the people of Dresden and Leipzig."[81]

In the West, too, people banked on what was tried and true, albeit in somewhat different form. Not less than twenty years after World War II ended, when the first German tank rolled off the production line again, the Federal Republic's Bundeswehr army would keep the Wehrmacht tradition going, once more choosing a beast of prey as the namesake for its new panzer tank.[82]

This time, the big cat was named "Leopard."

5

Raufbold

Das Anständigste bei der Jagd ist das Wild.
[What is most decent about hunting is the game.]
—Adolf Hitler, *Adolf Hitler: Monologe im Führerhauptquartier*

In the shade of the trees stands a moss-green colossus, looking as if he has just stepped out of the sheltering thicket and into the clearing. His head raised and crowned with a sweeping set of antlers, he gazes motionlessly into the distance, as if he were pausing to sniff whether the air is clear. And yet he does not move. He has been standing this way for a long while in Berlin's Tierpark. People and times have come and gone; patina now covers his metal coat. Nothing reveals what his name is or where he came from. Most everyone just passes by, paying no attention to him. And yet he would have much to tell them if he could.

His story began more than eight decades prior and around twenty kilometers further to the west. On a fall morning in 1937, the people flowing along Berlin's Masurenallee could already see his golden body from a distance, gleaming even in the murkily somber November light. From atop his granite pedestal he watched over the scene; without moving, he gazed up the boulevard to Adolf-Hitler-Platz, the square later renamed for the first president of the Federal Republic, Theodor Heuss. Hunters from all over the world had come to the German capital. There were French *par force* hunters standing around in their red jackets and white leather breeches, surrounded by a pack of hunting dogs excitedly jumping up and down. At the entry gate, a Finnish falconer with his bulky bearskin cap way down over his face sat astride his horse and was sullenly peering out from under it, while the golden eagle on his right arm looked agitatedly back and forth. To the left and the right of the golden deer statue, German foresters lined up next to one another. From the row of loden-green garb, only the hats with their *Gamsbärte* (literally, "chamois beards," the traditionally decorative tufts of goat hair), which resemble oversized shaving brushes, are conspicuous.[1]

The date was November 3, which was not just any date. It was St. Hubert's Day, a holiday that commemorated the patron saint of the hunt, who lived around 700 CE. According to the legend, St. Hubert—at the time still a young aristocratic daredevil—discovered a stag while out stalking prey, one that bore a glowing crucifix in his antlers. The sight so impressed Hubert that he allowed the stag to live; he became a pious Christian and later on even the bishop of Liège. Admittedly, the same story is likewise attributed to St. Eustachius, executed as an early Christian martyr around six hundred years earlier in Rome. It was only in the seventeenth century that in an attempt to lend the feudal hunt a Christian countenance, St. Hubert was made the protagonist of the legend. In this way, hunting was given a kind of moral justification—a pretense for venerating "the creator in the creature," as the hunting man of letters Oskar von Riesenthal described hunting in his 1880 poem "Waidmannsheil."[2]

On this particular St. Hubert's Day in 1937, however, Berlin did not quite commemorate past hunting saints, much less God's creatures. The huntsmen had come to Berlin to attend the opening of the International Hunting Exhibition. It was one of the few and final occasions on which the Nazis would not use the term "international" to disparage Jews but rather to delude foreign countries with the illusion of an open-minded, cosmopolitan Germany.[3] The exhibition was supposed to reinforce the "comradely cooperation among hunters across the entire world." At least that was what its patron would extol in retrospect as its greatest success.[4] It was, moreover, this patron who was the reason for the huge crowd that had gathered. At any moment Hermann Goering was supposed to appear.

Portly and Popular

Goering was the second most powerful man after Hitler and a National Socialist from the very start. He was there at the putsch attempted in Munich in 1923, had had a seat representing the Nazi Party in Berlin's national parliament since 1928, and, from 1932 on, worked vigorously in his capacity as the parliament's president to effect the failure of the Weimar Republic. Yet it was not so much his political convictions that led him to Hitler. The former fighter pilot—who had been awarded the Order of Merit, the highest military distinction, for shooting down numerous planes in

World War I—was looking for a new challenge in the turmoil of the young democracy. As described by journalist Joachim Fest, Goering was the archetype for a "'born' National Socialist," purportedly outfitted "with a spontaneous urge to prove [himself] in struggle and an unreflecting, elemental hunger for power."[5] Hitler, in turn, needed an ally who was as recognized as he was unscrupulous. Goering was substantially involved in setting up the Gestapo, had the first concentration camps built, and also was not afraid of having old fellow travelers killed if, like SA Chief Ernst Röhm, they stood in his way.[6] On the other hand, Goering's prestige as a war hero and his contacts in society's better circles helped make National Socialism *salonfähig* (socially respectable).[7] On top of that, his convivial, folksy manner made him one of the few leading Nazis who was approachable, which is why he was also at least as popular as Hitler among the people.[8]

At this time, toward the end of 1930s, Goering found himself at the height of his power. He was the minister-president of Prussia, president of the German parliament, Reich aviation minister, commander in chief of the Luftwaffe, and in charge of the four-year plan. In 1934, to be sure, he had to relinquish leadership of the Gestapo to Heinrich Himmler, but he had received two new appointments in return. After giving up the Gestapo post, he was able to call himself *Reichsforstmeister* (Reich forestry master) as well as—which presumably meant even more to him—*Reichsjägermeister* (Reich hunting master).[9]

The title *Reichsjägermeister* had first been used during the Holy Roman Empire under Emperor Maximilian, a historical fact that must have been entirely to Goering's liking.[10] After all, he had spent a large part of his childhood and teenage years at the castle of his godfather Hermann Epenstein that dated to the Middle Ages and was thought to be located on the site of an old Roman fort.[11] He loved the splendor of bygone eras and described himself as one of the last Renaissance men.[12]

Goering was always looking to make a grand entrance. Though the skies were cloudy that November morning and it was only barely six degrees Celsius, he pulled up in an open Mercedes. He was immediately recognizable by his immense girth. William C. Bullitt, a U.S. diplomat who visited him later that month, would subsequently say that Goering bore a strong similarity to "the hind end of an elephant."[13]

Goering not only enjoyed amassing offices and riches but also loved dressing himself up in all kinds of uniforms and would occasionally change his clothes up to five times a day.[14] He had the right outfit for every position. On this particular day, to go with his Tyrolean hat and its *Gamsbärte*, he was wearing an almost floor-length loden coat; each of the sleeves and collar points was studded with a golden stag's head, the so-called hart of St. Hubert. Instead of a crucifix between the points of its antlers, there was a gleaming golden swastika. It was the insignia of the Reichsbund Deutsche Jägerschaft (Reich League of German Hunters), which every German hunter had to be a member of.[15] Supposedly, Goering designed his uniform himself, as well as the league's green uniforms, patterned after the Luftwaffe's dress blues.[16] Thus fitted out and lined up like an honor guard, the German hunters stood there waiting for their *Reichsjägermeister*.

After Goering heaved himself out of his limousine, he marched along the rows, slackly signaling a Hitler salute, past the golden stag sculpture. The *Kronenhirsch* (crown stag) is what the Berlin sculptor Johannes Darsow called the statue, on account of its sweeping set of antlers. Darsow had designed and cast it specifically on commission from Goering for 1937's International Hunting Exhibition. The antlers of a red stag that Goering had shot down in early February 1936 served as the model for the statue. Yet he had abandoned Darsow's name and had been calling it "Raufbold" (Ruffian) instead.

The practice of giving stags with the most impressive antlers an illustrious name after they died had come into being during the reign of Kaiser Wilhelm II. After Goering was designated the supreme hunter in the Reich, they acquired such names even during their lifetime. Their names were supposed to be distinctive yet catchy but could not be repeated, and so over time, it became increasingly difficult to find names that had not been used yet. Most times the stags were named after a military rank, Germanic deities, or striking features and characteristics. They were called War Minister, Odin, One-Ear, or Ruffian.[17] At the exhibition, Raufbold's antlers hung in a room reserved for him in a special display that was created by *Reichsjägermeister* Goering. The only trophies that Goering put on display were those he believed warranted the public's marvel.

Goering had many of the deer he brought down immortalized in watercolors and oil paintings by Gerhard Löbenberg, his favorite hunting artist. Löbenberg indulged in that particular art motif that had arisen in the ateliers of nineteenth-century nature artists and that would subsequently settle into the parlors of the petite bourgeoisie, namely, the "bellowing stag." In the late nineteenth century, it became possible for the first time to produce art prints inexpensively and in large quantities, and these naturalistic images were popular among urban burghers. Thus the red stag became a mainstream motif. Ever since, it could be found hanging over dining room tables as well as nuptial beds, serving as an antipode to the industrialized city landscape or as a cipher for the pleasures of middle-class desire.[18] For the stag is a symbol of archaic virility, too, a sex egotist embodying "uninhibited promiscuity."[19]

Goering's painted stags do not bellow, though. Mostly they just stand there, with their head lowered or their majestic headdress flung backward. And yet the stags themselves seem to be only a decorative accessory, for it is always their antlers that constitute the focal point. Later on, Goering also had a portrait of Raufbold made. This stag, however, would be the only one for whom he would also erect a monument.

The attentive visitor to the exhibition would likely have noticed that there was a small yet subtle difference between Darsow's *Kronenhirsch* and Raufbold's remains. The trophy had a small flaw: on the right antler branch, the second point above the skull—the so-called *Eissprosse* (literally, "ice sprout")—is only faintly pronounced and barely longer than a thorn. Yet because this symbol for the International Hunting Exhibition had to be flawless, at Goering's behest the sculptor gave the statue two perfectly formed antler points.[20]

On the Hunt for a Ringleader

In retrospect, we can scarcely say when exactly and why Goering was seized by this hunting fever. Supposedly his mother had been an enthusiastic huntswoman and his godfather had frequently taken him along on hunts for chamois goats. Thus, because as a child Goering liked being out and about in the great outdoors, he presumably had first been exposed to hunting at a very young age.[21] If we are to believe Erich Gritzbach, his biographer and aide-de-camp, what he missed most during the *Kampfzeit* (time of struggle)—as the Nazis, in retrospect, called the less glamorous ups and downs of their party during the 1920s—was the time he had for hunting.[22] What is more likely, however, is that Goering had not been seized with this lust for larger and larger antler trophies until the beginning of the 1930s. Not least of all responsible for this was the hunting lobby.[23]

For centuries, hunting in Germany had been a privilege of the aristocracy. As bondsmen, peasants had no choice but to depend on princes to keep the deer and rabbits out of their fields. It was only in the wake of the

Revolution of 1848 that the German National Assembly in Frankfurt established that every property owner would be allowed to hunt on his own property and premises.[24] From then on, farmers were able to hunt for themselves and thereby protect their harvest and cover their own need for meat.[25] Soon enough, though, critical voices emerged that perceived "public order" and "common safety" to be in danger and that denigrated the farmers as "pothunters" without any sense of sport. In the years that followed, the freedom to hunt would be increasingly restricted by law, although, to be sure, it was still possible for farmers to drive wild animals from their land and, in uncertain cases, to bring that game down.[26]

Yet hunting organizations still wanted to curb hunting by farmers, for whom killing animals was a matter of providing meat and protecting their own harvest rather than securing trophies, and they sought to restrict the right to hunt to their elite circles again. When the Nazis seized power in January 1933, the hunting lobby perceived an opportunity to finally attain their long-pursued objectives.

They faced two problems in their efforts to assume control over who was able to hunt and when. First was the lack of a hunting law, which they had been demanding for decades. At the time of the Weimar Republic, regulations differed from one jurisdiction to the next. So in 1925, for instance, the red deer hunting season was closed for ten months in Bavaria, whereas there was no closed season at all in neighboring Hesse. Second, there were multiple hunting associations, which, moreover, quarreled among themselves. It was not until 1928 that the Reichsjagdbund (Reich Hunting League) created a joint umbrella association. Despite this step in the direction of greater organization, the majority of Germans who possessed hunting licenses were not members of any of these associations.[27] The hunters wanted to bring these individuals, too, under their control, a goal that accorded with the Nazi desire to control all arenas in life and organize them hierarchically. Among the new powers-that-be, then, the hunters sought out a "political advocate" for their interests.[28]

Why they hit upon Goering of all people is difficult to comprehend. He did not have any training in forestry or any experience as a hunter. But Social Democrat Otto Braun, Goering's predecessor as the Prussian minister-president and a longtime hunter himself, who had already made legal concessions to hunters in Prussia toward the end of the 1920s, had to flee to Switzerland as a consequence of the seizure of power. And Paul von Hindenburg, the geriatric Reich president, also a hunter, was presumably too old and weak for the hunting lobby. It is possible that they thought Goering could supply the requisite political clout. In retrospect, the German hunters maintained, in any case, that he was the best and only solution to their problem.[29]

Whatever may have convinced them that Goering was their man, going forward their lobby would put everything they had into making hunting palatable to him. They invited him on trophy hunts on the Schorfheide and along the Oder River. He was supposed "to taste blood"—and, as became evident rather quickly, their plan took off.[30] Already by May 1933, Goering was meeting with the leading hunting functionaries in Berlin to listen to their proposals for a new hunting law. On this occasion, apparently quite incidentally, they offered him the leadership of the German hunting world.[31]

For the impatient Goering, once he had been stoked, bringing the hunters under his direction could not happen fast enough. The hunting media were *gleichgeschaltet* (co-opted) and soon unanimously singing Goering's praises. The magazine *Wild und Hund* (Game and dog)—which had applauded Kaiser Wilhelm II for his mass game shoots and is to this day the hunting periodical with the highest circulation—celebrated the new "patron of German hunting" as a "very quick and very sure marksman."[32] The hunters were united in a new organization called the Reichsbund Deutsche Jägerschaft (Reich League of German Hunters). In July 1934— without the German parliament having been consulted on it—the Reichsjagdgesetz (Reich Hunting Act) was promulgated. In one stroke, it superseded the laws of seventeen different jurisdictions that had been in force until then.[33]

The new legislation fulfilled the wishes of the hunting lobby completely: it gave hunters their own honorary jurisdiction, enabling them to resolve hunting violations among themselves.[34] On top of that, only natural, not legal, persons were allowed to lease hunting grounds, which, for example, prevented any consortium of multiple hunters in a local community from doing so, making the latter once again dependent on large landowners.[35] The views of the hunters league also were a perfect match for the mindset of *Blut und Boden*. National Socialist *Rassenhygiene* accommodated the method of "Hege mit der Büchse" ("conservancy by rifle") very well.[36] Used since the turn of the century, the phrase owed to the Prussian *Forstmeister* (forestry master) Ferdinand von Raesfeld, who articulated this strategy for the first time in an article for *Wild und Hund* in 1895. The idea was to cultivate stags with the strongest and most sweeping set of antlers possible for trophy hunting and to kill weak animals with small or deformed antlers at an early stage. The deep accord between the hunting tradition and Nazi ideology was also voiced by Goering in a speech he delivered to the Reichsjagdgesetz:

> Love for nature and its creatures as well as the joy of stalking in field and forest is deeply rooted in the German people. Built upon age-old Germanic tradition, the noble art of the German hunter's craft thus developed over the course of centuries. . . . The duty of a proper hunter is not only to hunt game but also to conserve and care for it. . . . The right to hunt is inextricably connected to the right to the land on which the game lives and which feeds that game. . . . The trustee of German hunting is the Reich hunting master; he sees to it that no one carries a rifle who is not worthy of being the advocate for the heritage of the people entrusted to him.[37]

A *deutsche Waidgerechtigkeit* (German hunting code of ethics) was newly created by the Nazis, although what was precisely ethical and/or German about this code was never specified. Thus, for instance, it was considered "undeutsch" (un-German) to inflict any unnecessary pain on game animals. The use of buckshot or traps was branded as animal torture and forbidden. In this regard, admittedly, the well-being of the animal might not have been the reason for this regulation but rather the fact that these were the hunting methods of farmers, who were ultimately to be barred from hunting.

On the other hand, for instance, many a questionable hunting method was considered to be *waidgerecht* (ethical for the hunt) because it enjoyed a long tradition, such as the *Saufeder* (boar spear; literally, "sow feather").[38] Behind this harmless-sounding name hid a kind of lance that had been used since the Middle Ages, primarily in the so-called *Sauhatz* (the wild boar hunt). It consisted of an ashwood staff around two meters long, onto which a metal blade with a quillon was attached. Because wild boar are very aggressive and can reach speeds of up to fifty kilometers an hour, it was considered especially heroic to stand up to them while they ran at full gallop. Hunters embedded the *Saufeder* in the ground or propped it against a tree and then used dogs to goad the boar to them; they held the weapon out in front of them so that the fast-approaching animal would run straight into the blade. The quillon prevented the animal from getting too close to the hunter and injuring him. In the most advantageous case, the *Saufeder* pierced its heart or lungs, and the wild boar died immediately. Often, though, it bled to death slowly and tortuously.[39]

Goering placed more value on tradition than on animal welfare. He intended to create a set of *deutsches jagdliches Brauchtum* (German hunting customs), which would contain by what rituals the hunt was to proceed on German land and declare them to be "tradition." The first step in assembling this code was the compilation of all the customs that had developed in individual regions. Goering commissioned Walter Frevert to complete this task. Frevert was an experienced hunter and had been a member of both the Nazi Party and the SA since 1933. On top of that, since the end of 1936, he had been running the Nassawen Forestry Office. It was one of the four offices that East Prussia's Rominter Heath had been divided into.

Because in many places there were hardly any written sources about such customs, in many cases Frevert first had to invent them, such as, for instance, what dog breed was supposed to be used for boar hunting, how to ceremonially attend to the killing of a stag, and how to present all the game brought down in a hunt. In Saxony and Austria, the game that was killed had until recently been laid out on its left side after the hunt. Going forward, Frevert determined, it was to be laid on its right side, and the animals would be lined up left to right in descending order from the strongest to the weakest. Subsequently it would also be the custom to *verblasen* (trumpet; literally, "blast") the presented kill with hunting horns, and only the Fürst Pless horns from the province of Silesia and the half-moon bugles from the Sauerland region would be allowed to be used, as the larger French *trompes de chasse* that were quite common were considered "ungermanisch" (un-Germanic). French hunting melodies were forbidden as well, and the music for them was burned. Moreover, a *letzter Bissen* (final bite) was to be placed into the killed stag's *Äser* (mouth; literally, "grazer"). This often would turn out to be a twig from a fir tree, which the stag would never have eaten in real life.[40]

Whether these rituals made sense or not, Goering could not get enough of them.[41] He must have been more than satisfied with Frevert's diligent labors, for after Fervert completed this job, Goering entrusted him with special assignments over and over again. More often than not, Goering could not care less whether he himself complied with the legislation he enacted, as the fate that befell the red stag Raufbold, in particular, shows.

The Ogre of Rominten

In the early 1930s, Raufbold was living way out east in the German Empire, on the border with Poland, on the Rominter Heath. The forests there are the remnants of the so-called Grosse Wildniß (Great Wilderness), an impenetrable landscape of forests and bogs that covered vast parts of East Prussia through to the Middle Ages.[42] The region owed its name, Rominten (as it was also called), to the small Rominte River, which snakes its way lazily through the gently rolling hills there. Kings of Prussia had hunted along its banks, and toward the end of the nineteenth century, the heath was declared the hunting grounds of the imperial court. It was surrounded by a one-hundred-kilometer-long fence that included several entrances. These openings in the enclosure were mostly installed where the surrounding terrain was higher than the enclosed hunting grounds, so that wild animals could easily get into the reserve but were hardly able to jump out again.

The Rominter Heath comprised around twenty-five thousand hectares, which was approximately ten times the surface area of the Grunewald, the city forest in western Berlin. Shortly after the Nazis seized power in 1933, Goering reserved the heath for his own hunting parties and in the years that followed had the Rominten Reich hunting lodge built there. To this seigneurial estate, which was made to look like a log house lodge and had a thatched roof, he would invite only select hunting guests. Whenever he was able to break away from the responsibilities of his numerous offices—and that was more frequent than we might assume—he would go hunting there.[43]

The red stags of Rominten were renowned for their tremendous sets of antlers. Antlers consist of osseous matter that the stag sheds in the first part of every year, after which it subsequently grows a new and typically even larger set. How expansive the set becomes over the course of the stag's life is a matter of genetics. Environmental factors only play a role in how much the antlers weigh—their density, which is the decisive criterion for trophy hunters.[44] For that reason, Raufbold and his fellow red deer would be fed with oats, bran, and sesame cakes once winter set in.[45] That was how Raufbold developed into a massive red stag, a "twenty-pointer," as designated by the number of his antler branches. In the wide-ranging terrain of the heath he wanted for nothing. Moreover, there were hardly any natural predatory enemies lying in wait for him—the last bear had been killed 150 years before, the last lynx around 80 years ago.[46] Only starting in 1940 did Goering have individual lynx deliberately reintroduced.[47] Sometimes wolves from Poland searching for food would roam through the Rominter Heath. Yet they would soon disappear again, since Goering's forestry officers only too gladly hunted them down.[48] In the long term, there was only a single predator that ruled in Rominten—Goering himself. In April 1938, he declared Rominten to be a state hunting reserve and removed it from the auspices of general forestry administration so that from then on he would be the absolute ruler there.[49] Around three decades later, the French writer Michel Tournier picked up on the image of Goering as the sovereign of the heath in his novel *Le roi des aulnes* (*The Erl-King*), describing Goering as a mythical, man-eating figure, the "ogre of Rominten."[50]

The ogre Goering was out to get the massive stags and their antlers. Over time he went from being a trigger-happy stalker to being a deliberate trophy hunter. He would only shoot down stags that his forestry officers described and deliberately selected for him beforehand, those that he knew were at the peak of their lives.[51] Most red deer reach that apex with their "neunten Kopf" ("ninth head") in hunters' lingo, that is, at the age of ten. Then their body is fully grown and they have sufficient energy left over for growing a set of antlers. After their twelfth year, both the antlers' weight and number of points gradually decrease again.[52] The stag "resets" at that point, according to German hunters. Therefore, their goal is to catch the stag before that point, right at that time when he is *reif* (mature; literally, "ripe"). Every summer in Rominten, in order to determine this moment, a so-called *Stangenparade* (antler beam review) was conducted, during which the reserve's foresters would examine the sets of antler beams that stags on the heath had shed in late winter and then establish which stags would be released for killing in the coming fall.[53]

How old Raufbold was when Goering shot him we can only guess. He was probably born at the beginning or in the middle of the 1920s. The only thing documented for certain was the day of his death, February 9, 1936—a few days later and Raufbold would perhaps have already lost his antlers.[54] What would have remained would have been only the shed antler beams, left behind in the forest, where mice and other rodents would have dug into them. Formerly, the month of February was also called *Hornung* in German, because it is the month when red deer shed their antlers. So time was of the essence. Goering did not want Raufbold's set of antlers to elude him, nor did he want to wait yet another year only maybe to find out that the antlers were not that impressive anymore. For him, then, it made no difference that the six-month closed season mandated by law had begun on February 1.[55] All that mattered to him was the trophy.

Weighing eight kilos, Raufbold's antlers even brought him a medal of the first class at the International Hunting Exhibition—though they were far from being the exhibition's heaviest trophy.[56] Goering did not try to hide the fact that he did as he pleased, his law be damned. Any visitor could see that the *Führer der grünen Gilde* (leader of the green guild) had broken his own law in order to bring down Raufbold, as the date he was shot was recorded on the trophy.[57]

Grand and Gross

On the third day of the expo, even Adolf Hitler finally visited the exhibition halls. With his arms crossed and a disgusted look on his face, he went past the seemingly endless rows of bones and hides.

Hitler did not have a high opinion of hunting. For him it was nothing more than a "feiger Sport" ("cowardly sport").[58] Every once in a while, in familiar circles, he would let loose about Goering's passion: "If only there were still some danger connected with hunting, as in the days when men used spears for killing game," as he said during one of his rounds of talks in his headquarters. "But today, when anybody with a fat belly can safely shoot the animal down from a distance . . ." For him, hunting was like horse racing, one of the last "remnants of a dead feudal world."[59] Only for poachers did Hitler harbor certain romantic sympathies, because in his opinion they supposedly still risked their lives while hunting.[60] Indeed, in January 1940, he decreed that poaching was not grounds for being barred from the party.[61]

He was not alone in his dismissive stance toward hunting. Propaganda Minister Joseph Goebbels considered the new hunting legislation and the whole hunting business to be fundamentally reactionary.[62] Martin Bormann, Hitler's *Kanzleileiter* (chancellery leader), even complained that many *Gauleiter* (regional district leaders) were no longer devoted to anything but "their cursed hunting."[63] Even if he described hunters as "green freemasons," Hitler still did his "best man" and "most loyal paladin" the favor, albeit disgruntledly, of entering his name into the "green book" of the International Hunting Exhibition.[64]

In Hitler's circles, Goering was controversial on account of his decadent lifestyle and narcissistic behavior. Goebbels, who had a penchant for recherché terminology, disparagingly called him a "sybarite," a hedonist.[65] Above all else, though, the propaganda minister was disturbed by Goering's "lack of firmly held convictions." Goebbels remarked that Goering supposedly had as much to do with the party "as a cow does with radiology."[66] Being met with such hostility did not bother Goering at all. He made no secret either of the fact that he was not driven by idealism or any sort of political agenda but rather solely and exclusively by the desire for prestige and the hunger for power.

Around two years after the International Hunting Exhibition, the *Kronenhirsch* statue moved to a new spot at Goering's estate on the Schorfheide, a forested area of more than one thousand square kilometers that is about forty kilometers north of Berlin. As with Rominten in East Prussia, it, too, had been a former hunting reserve of Prussian kings. There, on a 120-hectare plot of land between two lakes, sat Carinhall. From the outside, it looked like a Swedish log house, except for the terrace door that was decorated with swastikas. Inside, however, it exuded the grandeur of a fairy-tale castle, and the walls were entirely covered with trophies. When the Yugoslav prince regent Paul Karadjordjevic visited Carinhall, he was so impressed by its splendor that he cried out in amazement that "not even the czars had anything like this!"[67]

The value of the gigantic estate was estimated at more than eighteen million Reichsmarks—almost seventy million in today's euros.[68] Goering did not have to pay from his own pocket for either the construction or the numerous expansions; instead, it was paid by the Prussian government and the Reichsluftfahrtministerium. For Carinhall had one purpose above all else: it was where Goering was to represent the government and the party to best effect.[69]

He had chosen the name Carinhall in memory of his deceased first wife. On the shore of the lake he had even erected a crypt specially for her and had had her reinterred there from her grave in Sweden, a relocation that had caused a lot of brouhaha. At Carinhall, Goering in his capacity as *Nebenaußenminister* (adjutant foreign minister) primarily received guests of state he had befriended, like the Italian duce Benito Mussolini and the British ambassador Neville Henderson.[70]

The road into Goering's realm went past two small masonry guard-houses, behind which was a sandy allée lined with horse chestnut trees that ran straight through the woods for about one kilometer.[71] At the western end of the lane, shortly before the courtyard, visitors would pass the so-called Stag Square. The gigantic sculpture of the stag was visible from far away. It was Raufbold, cast in bronze, gazing out across the clearing from his pedestal. In the years that followed, countless guests would pass by him. And he would remain there long after Goering himself had left the Schorfheide reserve.[72]

For the time being, however, Goering was as yet the absolute ruler on the Schorfheide. Not far from his palatial residence, a wooden stele was erected in his honor that showed the extent of his vanity and the hunts-men's bondage. The inscription read:

To the Reich hunting master Hermann Goering
Our thanks for the new hunting act
The wild animals of Germany[73]

Admittedly, Goering did not invite only presidents, diplomats, and aristo-crats to the Schorfheide. In June 1939, for instance, on the occasion of the meeting for the Deutscher Forstverein (German Forest Union), he hosted a party on the western shore of the nearby lake called Werbellinsee. He had forty travel coaches bring the two thousand foresters from Berlin, where they had gathered from all over the Reich. The view they were offered must have rendered them speechless. Any number of guests no doubt recalled the lines by writer Theodor Fontane, who once whiled away his time not far away on the shores of the lake: "It is an enchanted spot where we sit, for we're sitting on the shore of Werbellin."[74] The self-proclaimed "Renaissance man" Goering spared no expense. Whole oxen and hogs were roasted on spits above gigantic bonfires. While Goering shot at wooden mock-ups of wisent with a bow and arrow, the city ballet of Berlin danced on a float that drifted by on the lake.[75] Alcohol flowed freely and abundantly. In addition, we can assume that besides the beer and wine found on the long wooden tables, there were also bottles of a particular brown alcoholic drink that owed its popularity in large measure to Goering and his green following as well.

It had begun with an idea for schnapps conceived by a vinegar producer and wine dealer from the town of Wolfenbüttel in Lower Saxony. In 1934,

Curt Mast came up with the idea to expand his assorted offerings with spirits and bring the first industrially produced herbal liqueur onto the market as a way to save his crisis-hit family operation from ruin. He was missing only a suitable name. It had to be one that would immediately stick. Because Mast liked hunting in his free time, he knew that the 180,000 hunters in Germany—who, after blowing the *Halali* trumpet signal, also "tot tranken" ("drank to the death") of every animal they killed—were the perfect target demographic.[76]

Initially, Mast thought he might call the herbal schnapps "Hubertusbitter" ("St. Hubert's bitters"). Then, however, he remembered that Goering had been operating under the title of *Reichsjägermeister* and that, furthermore, he had organized the hunting administration hierarchically, creating the positions of *Oberstjägermeister* (supreme hunting master) and *Gaujägermeister* (regional hunting masters) and *Kreisjägermeister* (district hunting masters). The name "Jägermeister" pleased Mast so much that by the end of March 1935, he had patented it as a trademark and had commissioned a graphic artist to design the label. The logo with the hart of St. Hubert has been preserved in its essence to this day. The drink went on the market in the same year and would soon meet with success—and not only among hunters.[77]

In Pursuit of the Brute

Parties like the 1939 fest on the Schorfheide also show how hunting was not just a private pastime or an end in itself. Above all, hunting made it possible for Goering to cultivate his contacts on a comradely level, although he was not nearly as generous as he made himself out to be at festivities, especially when it involved deer. His trophy envy was pronounced, and he could hardly bear anyone else bagging a stag under his nose. In September 1935, that envy nearly ended delicate relations with Hungary even before they had properly begun. During a joint hunting party in Rominten, the Hungarian minister-president Gyula Gömbös shot a stag that had been intended for Goering. Goering was beside himself with rage and only calmed down thanks to Gömbös's easy-going nature. Subsequently, Goering swung to the other extreme and, as a sign of his benevolence, had a nearby lake named after Gömbös.[78]

With Joachim von Ribbentrop, Goering would not be quite so charitable. The German ambassador in London was appointed as foreign minister by

Hitler in February 1938. As a hoarder of offices, Goering would have gladly occupied this post himself, not to mention that he could not stand von Ribbentrop. When he also found out that von Ribbentrop, without his permission, had shot an "outstandingly promising stag" in Rominten—one from whom Goering had hoped to score an even more stately set of antlers for himself in the years to come—he promptly saw to it that von Ribbentrop would never be allowed to set foot on his hunting grounds again.[79]

Once in a while Goering did get to feel the limits of his influence, however. Three years after the incident with von Ribbentrop, in September 1941, *Generalfeldmarschall* (general field marshal) Walther von Brauchitsch brought down the stag called Eggenhirsch (Harrow-Hart), an *ungeraden Dreißigender* (uneven thirty-pointer)—that is, a beast that has one fifteen-point antler beam and one fourteen-point antler beam. Von Brauchitsch's booty counted among the most powerful red stags that had ever been shot there. Frevert, the *Oberforstmeister* (chief forestry officer) of Rominter Heath responsible for Goering's deer-hunting guest list, had selected the Eggenhirsch specifically for von Brauchitsch. Initially, Goering let nothing show and marked the occasion with champagne, that is, until he saw the presentation from the hunt. There, in accordance with the customs of the hunt, Eggenhirsch lay all the way to the left and the two weaker deer shot by Goering to the right of it. Then he yelled at Frevert: "You, sir, allow my guests to shoot the twelve-point stags and larger, yet leave me personally with your abnormalities!"

"But, Herr Reichsjägermeister," Frevert responded, "this guest is also commander in chief of the military," whereupon Goering made a muttering retreat.[80] A few days later, he issued the following decree to all state hunting reserves: "In future, twelve-point stags and larger on the reserves where I am accustomed to being personally present during their rutting season will be brought down only by me and not by my guests. . . . Exceptions . . . can be ordered only by me."[81]

Soon enough, Goering would not be satisfied with deer alone. He dreamed of the hunt in a primeval wilderness, filled with the game that the ancient Germanic tribes hunted and that had found its way into the Reichsjagdgesetz. For the legislation not only mentioned the *jagdbar* (huntable) game species of moose, wisent, and Alpine ibex but also those that were long extinct in this part of the world. That was about to change, however, because, fortunately for Goering, there was Lutz Heck.

Since the 1920s, this director of the Zoological Garden in Berlin, along with other colleagues, had been working to ensure the preservation and repropagation of the last remaining wisent living in captivity.[82] Goering was so enthusiastic about such efforts that he had several of these European bison put out into an enclosure on the Schorfheide. However, Heck's most sensational undertaking was de-extinction. Together with brother Heinz, the director of the Hellabrunn Zoo in Munich, he would attempt to resurrect the aurochs.

Numerous legends had sprung up about these brutish bovines, the largest of the wild cattle in Europe, which had been celebrated as the ur beast in the *Nibelungenlied* and which reached a height of up to two meters at its withers, even surpassing the wisent. Yet on account of its meat, its hide, and, above all, its horns—which in the bulls could reach a length of more than a meter—it had been hunted so intensively that it was already largely extinct by the thirteenth century. It could still be found in East Prussia at the beginning of the sixteenth century, though it disappeared shortly afterward there as well. In the end, the last beast of its sort died in 1627, in the vicinity of Warsaw, presumably from old age.[83]

To bring the aurochs back to life, the brothers Heck crossed diverse domesticated cattle breeds with one another until their offspring started to look like the wild version again. For that reason, they were not uncontroversial among zoologists. Many of their colleagues criticized their attempts as unscientific and disparagingly called them "Urmacher" (referring to their role both as "aurochs makers" and as pursuers of the primeval).[84] Goering, though, was excited about the breeding attempts, because they brought him the promise of new, extraordinary hunting trophies. In 1938, as thanks, he would entrust Lutz Heck with running the nature conservancy division at the Reichsforstamt (Reich Forestry Office), and on April 20 of that same year—on the *Führergeburtstag* (the führer's birthday)—he awarded him the title of professor.[85]

Beginning in 1938, Goering had a small herd of Heck brothers' cattle established in Rominten, where, however, they soon caused a commotion.[86] For the Hecks' back-bred bovines not only looked like their extinct ancestral ur cow but also behaved like them. The alpha bull, for instance, stole feed from the horses of local forest workers and then went after the men, while one of the cows attacked a group of hikers. Only after it was determined that the cattle were driving the red deer from their feed stations, however, were they recaptured and resettled.[87] Nature conservancy for Goering meant, first of all, securing his hunting grounds. And his stags meant more to him than anything.[88]

Under German Trees

Although as subspecies, red deer range across all of Europe from Scandinavia to the Mediterranean, during the Third Reich they were still only referred to as the "deutscher Edelhirsch" ("German royal deer"). For when the forest itself is "German," it seems, then the "king of the forest" must be a German, too. As Elias Canetti notes in his magnum opus *Crowds and Power* (originally published in German as *Masse und Macht* in 1960), "In no other modern country has the forest-feeling remained as alive as it has in Germany."[89]

The myth of the German forest has a long history. It starts with the Cherusci and their chieftain Arminius (Hermann the German), who vanquished the Roman commander Varus in the Teutoburg Forest in 9 CE. Among the oldest preserved sources for the storied slaughter is the monograph *Germania*, which the Roman historian Tacitus composed around a hundred years after the event.[90] The precise site of the bloody battle has been the subject of much speculation. Presumably it was located in the vicinity of Osnabrück, in present-day North Rhine-Westphalia.[91]

For the nascent national pride of the late eighteenth century, however, exactly where it happened made no difference. Poets and playwrights like Friedrich Gottlieb Klopstock, Heinrich von Kleist, and Joseph von Eichendorff as well as painters like Caspar David Friedrich took up the myth in their works.[92] It was in the forest—their traditional *Lebensraum*, as legend would have it from that point forward—that the Germanic tribes overcame the enemy conqueror. Arminius became Hermann—the first German hero.

In this era, ideas about the forest changed too. Though for centuries the woods had been considered a den of danger where wild beasts and bandit gangs hid out, during the Romantic period the forest increasingly lost its threatening character and became the symbol of Germany, its unique, distinguishing feature. The historian and poet Ernst Moritz Arndt even saw a close connection between the landscape and the *Volkscharakter* (national character).[93] Wilhelm Heinrich Riehl, an ethnographer and student of Arndt, further developed this idea and attempted to deduce from it the status of individual nations. In the freely accessible forests, he discovered the central feature distinguishing Germany from other countries, like England, which for the most part had only "enclosed parks" instead of open forests. And, according to him, "in this freedom of the German forest, which peers out so strangely from our other modern institutions, there are more determining influences on our highly cultivated lives—namely, on the romantic sentiment in them—than many could ever dream of."[94] Therefore, he said, the goal had to be to preserve the forest unconditionally, "not just so that the ovens do not go cold on us in the winter, but also so that the pulse of the national life continues to beat warmly and merrily, so that Germany remains German."[95]

The German forest became the symbol for the *Heimat* (homeland), which like it had to be defended. After World War I, *völkisch* movements increasingly made the forest their own, so as to distinguish themselves from certain other *Volksgruppen* (national groups, or ethnicities). In the process, they designated the mixed deciduous forest as their natural ideal, dismissing the monoculture of the pine forest as a symbol of "communist egalitarian claptrap."[96] As Walther Schoenichen, the longtime leader of the Staatliche Stelle für Naturdenkmalpflege in Preußen (State Department for the Protection of Nature in Prussia), noted in 1934, it was only "in the harsh struggle with the forest that the German *Mensch* created his *Lebensraum* with dogged determination."[97]

In a 1938 pamphlet titled *So lebt die Waldgemeinschaft* (How the forest community lives), the Nazis explained how the "natural composition of the forest" conveyed the established hierarchy of the *Volksgemeinschaft* (national community). The pamphlet was intended to teach children about the commonalties between the forest and *Volksgemeinschaft* through the use of diagrams that represented the tree canopy overstory as making up the dominant stratum, the understory and shrubs as constituting the middle stratum, and the grasses, mosses, and the forest floor underneath that made up the herbaceous and ground cover layer as forming the lower, dominated stratum: "The topmost stratum is composed by a few. They can be counted (and count). The further down you go, the more these comrades in life are combined in one stratum."[98]

To reach the wider population as well, in 1935 the NS-Kulturgemeinde (Nazi cultural community) shot the black-and-white film *Ewiger Wald* (Eternal forest) at the behest of Alfred Rosenberg, the head ideologue of the Nazi Party. The film was supposed to convey the National Socialist

Weltanschauung, packaged as art for the people.[99] In the film, shots of forests and trees alternate with feature film sequences covering historical topics ranging from the Stone Age to the Battle of the Teutoburg Forest, the *Ostsiedlung* (settlement of the east) by the Teutonic knights in the High Middle Ages, and the pastoral peasant idyll of the Nazi era.[100] Voice-overs of chanting and a male voice that called to mind with mantra-like verses the "everlasting nature of the fate shared between the German forest and the German people" accompanied the images.[101]

The public response to the film, which arrived in movie theaters in the summer of 1936, was modest. The moviegoers were baffled by lines like "We come from the forest / We live like the forest / Out of the forest we make / Home and space." Even the reviews in the co-opted press were lukewarm. The party had apparently anticipated this reaction, for it had released the film only under the condition "that neither the film nor announcements" for the film reference it.[102] Rosenberg supposed that the propaganda minister Goebbels was responsible for the condition, although Goebbels did not even know the film's name (in his diary, he called it *Der deutsche Wald* [The German forest]), and according to him, the condition came from Hitler himself.[103]

Hitler loved to go on automobile drives through the forest "far away from the throng"; he had this view of the forest because, as he saw it, only "idiots" lived there, not civilized human beings.[104] And he had little regard for the theory of heroic forest peoples. "We have a false picture of the battle of the Teutoberg [*sic*] Forest," he told his listeners at his headquarters in the middle of August 1941. "The romanticism of our teachers of history has played its part in that. At that period, it was not in fact possible, any more than today, to fight a battle in a forest."[105] Only "subjugated peoples," in Hitler's judgment, would retreat back into the forests.[106]

Yet the German forest was more than just the site of Goering's romanticized ideas about hunting and nature. It was also first and foremost a crucial supplier of raw materials. "We have to get as much as possible out of the German forest," Goering said in his first session with the administrators for the state forestry service on July 11, 1934. To be sure, he also promoted the establishment of nature conservation areas. In order to protect the homeland, it was not enough to support "the business of lumber production"; "primeval stands of timber should also be maintained and cared for as a refuge for highly valued wildlife."[107] Protecting nature, however, was ancillary. Preparations for war had long been in full swing, and Goering was responsible for the four-year plan. Under Goering, the forest was far from preserved. In fact, twice as much timber was felled as sustainable use would allow. By the end of the war in May 1945, there would be 14 percent less forest than there was when the four-year plan was launched in 1936.[108]

With the invasion of Poland in September 1939, the forest acquired yet another meaning as new *Lebensraum im Osten*, a new homeland in the east. The so-called *Generalplan Ost* (general plan for the east) envisioned, among other things, transforming the landscape along the lines of the German model so that the conquered areas would provide the new settlers with "surroundings that suited their German soul."[109] The Reichsforstamt therefore pursued a "reforestation of the east."[110] Once again there was talk of the "German peoples of the forest," who were contrasted with the "Slavic peoples of the steppe" and the "Jewish peoples of the desert."[111]

Even if Goering was initially skeptical about a new war, he still used the conquests for the purely personal purposes of expanding his hunting reserves and acquiring new ones. In the fall of 1939, he enlarged the Rominter Heath and added a fifth forestry office. The land that made this expansion possible had until that time had been located in Polish territory. Ten villages had to yield to Goering's plan, and the inhabitants were expelled to the occupied parts of Poland.[112]

Manhunting

Two years earlier, at the International Hunting Exhibition, in November 1937, Goering had made it clear that he was also out to get other hunting areas in Poland.

When he visited the exhibition halls several days before the opening, he could hardly get enough of the Polish collection. There, hanging side by side, were wolves' hides, lynx pelts, and red stag antlers. On the platforms beneath them were row upon row of mounted birds beside other deer antlers. Amid all these cadavers and bones reigned a stuffed wisent. What sparked Goering's interest the most, though, was a relief map the size of a dining table. It showed Białowieża, the last primeval forest in Europe where the last wild wisent were living.

Goering was familiar with the area. He had been invited to hunt there by Józef Lipski, the Polish ambassador in Berlin, in March 1934 and was so enthralled that he returned every year from then on.[113] Raving about the primeval natural landscape and its abundant stocks of game, Goering stood there with his entourage in front of the map. Putting his left hand casually into the pocket of his trousers, in his right hand he held a riding crop that he, seemingly in passing, swept over the country stretched out before him, like a field commander who was planning his next strategic campaign.

A photo of this scene was included in the exhibition catalog. Yet the publishers must have been conscious of the unmistakable symbolic power of the image of him holding the whip, which explains why they airbrushed it out of his hand.[114] At that point, two years before the attack on Poland, such a belligerent gesture would not have fit with the image of itself that Nazi Germany was aiming to present. Goering was still making a splash as Hitler's most important diplomat. Yet it was only four years later that he would turn his dream of a "primeval Germanic hunting forest" into reality.[115]

After Germany and the Soviet Union overran Poland and split the occupied country up between each other, Białowieża was occupied by the Russians. That changed, however, in the summer of 1941, when Germany invaded the Soviet Union. Then, nothing really stood in Goering's way anymore.

Goering wanted to establish a gigantic *Reichsjagdgebiet* (Reich hunting region) there and to enlarge the forest from sixteen hundred to twenty-six hundred square kilometers, an area that would be larger than the entire metropolitan area of Berlin. Goering assigned Walter Frevert with the task of making this expansion happen, naming him as head of the newly created, now Germanized Bialowies Oberforstamt (Białowieża Superior Forestry Office) and, on top of that, giving him a special assignment. He was to "bring peace" to the forest and "evacuate" it.[116]

Frevert, who was already administering the Rominter Heath as *Oberforstmeister*, was an unscrupulous pragmatist.[117] If it served his career as a forester and his passion for the hunt, then he was prepared to do anything. Many years later, he would say, "For Rominten I would have signed a pact with the devil himself!"—and not just for Rominten, as he would prove in Białowieża.[118]

A group of a hundred men from the Forstschutzkorps (Forestry Protection Corps) was put at Frevert's disposal. These paramilitary units had been deployed in occupied Poland since 1940. They were recruited from armed forest workers and foresters and were originally supposed to secure the transport of lumber as well as engage in combat poaching and illegal arms trading in the forested areas.[119] In Białowieża, however, they had another assignment. Starting in July 1941, Frevert's men combed through the forest to track down scattered soldiers from the Red Army. Next, they bore down on the local population, informing them that they would have to chop down fewer numbers of trees than they had been accustomed to. At the forest's edge, they closed the industrial shops where charcoal was made and lumber milled. Then, at the end of July 1941, Frevert's unit together with the 322nd Police Battalion, which had been relocated there specifically for this assignment, began to clear the Bialowies Reichsjagdgebiet of people.

In the early morning, they burst into the villages and turned the houses inside out. They gave the residents half an hour to pack up what they needed most. Before Frevert's men set the houses on fire, they took what they could use themselves. Within one week, almost seven thousand people were displaced and thirty-four villages burnt down.[120]

The police division was even more ruthless toward the Jewish population. At the start of August 1941—around two months before the mass deportations of Jews from Germany would begin—the women and children in the vicinity were taken to the ghetto of Kobryn, ninety kilometers away. The 584 boys and men were shot to death on the spot.[121] A few managed to escape into the woods, where the Forstschutzkorps, the police, and a *Sonderkommando* (special task force), specifically deployed for the purpose by the Luftwaffe, would hunt them from then on.[122] The expedition was officially called "Bandenbekämpfung" ("combatting banditry").

In the fall of 1941, Frevert left Białowieża for Rominten, where Goering was already expecting him, as the deer's rutting season was about to start. The next two years, though, he would return multiple times at Goering's behest to the Bialowies Reichsjagdgebiet to engage in *Bandenbekämpfung*. As he wrote to a hunting buddy in March 1942, "there are still partisans and other bandits here in great numbers, and the bounty for them is considerably larger than for any game."[123]

End Points of the Deer

While people were being hunted on behalf of Goering in Białowieża, in his reserve in Rominten he concentrated on what he saw as more essential, namely, deer hunting. His greed for larger and larger trophies knew no bounds. For he was pursuing a goal that unites many hunters—the *Lebens-hirsch* (stag of a lifetime). That is what they call the most powerful deer they try their whole lives to bring down, the one that is to be the crowning glory of their hunting quests.

On average, Goering shot five stags every year, ever since he had started hunting in Rominten. Numerous sets of antlers on the walls of the *Reichs-jägerhof* testified to his exploits. Many of them surpassed Raufbold, the symbol for the 1937 International Hunting Exhibition, by a long shot. Yet he had not ever brought down one like Matador before, a stag he had observed at a feed station through his binoculars in November 1940, while his Luftwaffe made one airstrike after the other on English towns like Coventry.[124] His forestry officers reported to him that its antlers gained around two kilos in weight year after year. In the fall of 1942, Matador was a twenty-two-pointer, estimated to be about nine years old. As a rule, the antlers of Rominten stags hardly ever became heavier after the animal reached the age of ten but rather began losing weight instead. Time was therefore of the essence. The risk was too great that either the stag would lose its life during a battle over turf—although that rarely happened—or that another hunter would nab him, a prospect Goering could not countenance. In fact, Matador had nearly been shot before by a friend of Frevert's, when the stag once lost its way in the woods of another forestry district. Luckily, though, Goering came to know nothing of that near miss.[125]

On September 22, 1942, he finally brought down Matador. Measuring more than a meter, each of its antler beams weighed almost twelve kilos. It was the heaviest trophy that had ever been bagged in Rominten; even Kaiser Wilhelm II, who had been the supreme commander of the hunting reserve before Goering, had never shot a more powerful stag. For two hundred years, there had only been two stags in all of Europe that were stronger.[126] Matador was Goering's *Lebenshirsch* in the truest sense of the word. From here on in, it was downhill for Goering.

His Luftwaffe would score increasingly fewer wins. Before the war, Goering had supposedly boasted that he wanted to be called "Meier," an ordinary German surname with undertones similar to Smith or Jones in English, "should even one single enemy airplane fly across the German border."[127] The battle for the skies over England had long been lost; the Allied airstrikes on Germany had been an everyday experience for months.

In the air raid shelters underground, from Cologne to Königsberg (now Kaliningrad), people tried to distract themselves with jokes about "Hermann Meier."

Goering had Carinhall built into a fortress in the meantime. In 1942, the thatched roof was traded in for tiles; the buildings were draped with camouflage netting; and the number of flak guns increased from four to nine. That was still not enough, though. A few kilometers further to the north stood a precisely detailed re-creation of Goering's royal residence. Only a few older *Pioniersoldaten* (volunteer home guard soldiers) were holed up there; in case of an air raid warning, they used pyrotechnics and lighting to make the wood-and-canvas reproduction seem like the real Carinhall.[128]

With this security, Goering found it bearable. When Josef Terboven, the *Reichskommissar* (Reich commissioner) for occupied Norway, was visiting him one Sunday, even though enemy units were reported all over Germany, Goering merely had himself briefed by his adjutant as to whether an air raid warning had been issued for Carinhall. When the aide-de-camp gave him the all clear, Goering breathed a sigh of relief and said contentedly: "Fine, let's do some hunting, sir."[129]

But his obsession with hunting was having an impact on his own people too. In the winter of 1942, it led to the so-called *Haferkrieg* (oats war). While the 6th Army was bottled up near Stalingrad and could barely still be supplied and even homeland provisions were running out, the deer on Goering's state hunting reserve were still being fed with oats. Since they could have been used to feed small children, complaints were lodged from multiple *Ernährungsämter* (food supply bureaus) and *Gauleiter*. Because the war morale of the population was not to be jeopardized, in the end Goering's Reichsjagdamt had to abandon the feedings.[130] No doubt Goering also conceded on this point because Hitler's manager Martin Bormann came to know about it and remarked that he was quite worried what would happen when the führer learned about it.[131]

Hitler had made no secret of how pitiful he found hunting to be. He had looked the other way, so long as Goering was scoring wins and presenting the Nazi government in a good light to the outside world. Yet Goering had been falling from Hitler's good graces since at least 1938, for Hitler believed that Goering objected to the war only because he was

concerned about his ability to indulge in his accumulated riches and priv-
ileges with a war going on.[132]

During this phase, Goering seemed more and more like a stag during
fattening season, those weeks in late summer when stags retreat and avoid
any fighting, ravenously putting on fat reserves for the rut that will sap all
their strength. As one saying among German hunters goes, "Der Feisthirsch
ist ein Waldgespenst, das du nur ahnst und niemals kennst" ("The fattening stag is a forest ghost, that thou but sense'st and never know'st"). The
greater the political setbacks Goering faced the more he retreated. He only
appeared rarely for briefings at the führer's headquarters. Moreover, when
he did, the scene typically played out as it did in August 1944.

Wearing a paratrooper's uniform and hunting boots, Goering stormed
into the briefing room where Hitler was seated at a table. He was concerned
about the Russian advance. Even as people in East Prussia were emptying
their bank accounts, the trains heading westward were overflowing, and
endlessly trekking trails of people were fleeing the approaching Red Army,
Goering was only worried about his game: "My poor deer. It's horrible!"[133]
Shortly after this briefing, rutting season began in the Rominter Heath.
The familiar grunting of the stags mixed more and more with the faraway
cannon thunder. The Allied front edged closer. There were only ten kilometers left between it and boundary of the heath. At the beginning of
October, after the first Russian paratroopers landed in Rominten, Goering
and his loyal lackey Frevert would go hunting one last time, but this time
Goering was unsuccessful. He left his reserve the following day, never to
set foot on the heath again.[134]

So that his hunting estate and his accumulated riches would not fall
into Soviet hands, Goering had everything that was valuable taken away
in special trains. Even the guard staff and forestry officials left the heath.
In the end, only a caretaker stayed behind in order to complete the final
act of Goering's reign, which bore the cover name of "Johannisfeuer" ("St.
John's Bonfire").

For ages, it had been a Christian custom to light a bonfire on St. John's
Eve at the end of June, so as to keep evil spirits at bay. Goering's bonfire,
however, would serve only one purpose, namely, to ensure that nothing
but scorched earth remain. On October 20, 1944—almost two weeks after
Goering left Rominten forever and shortly before the Red Army reached
the estate—the *Reichsjägerhof* went up in flames.[135]

In the spring of 1945, Joseph Goebbels read in the newspaper *Joachimsthaler Zeitung* that Goering had shot one of his precious wisent on the Schorfheide so as to supply rationed meat for the fleeing population. Goebbels was infuriated. For him, as he wrote in his diary, this action represented what "more or less demonstrates the height of degeneracy reached by Goering and his entourage." He forwarded the newspaper article to Hitler with a note that Goering reminded him "of the Bourbon princess who, as the mob stormed the Tuileries shouting 'Bread!' asked the naïve question: 'Why don't the people eat cake?'"[136]

Goering issued the orders to blow up the premises even before the Red Army had reached Carinhall on the Schorfheide. He himself, along with his family, absconded first to his summer home in Bavaria and shortly afterward to the Salzburg region, where the Allies would take him prisoner in May 1945.

When the Soviet troops marched up the boulevard of chestnut trees leading to Carinhall, they could already see from a distance that nothing but rubble was left of Goering's realm. Yet amid all the debris, forlorn on its pedestal, an upright bronze stag stood watch. The *Kronenhirsch*. Miraculously, it had withstood the detonations. The Red Army took it away and put it first in the garden of an occupied villa in the Babelsberg district of Potsdam. Then, in the 1950s, the sculpture moved to the park at Sanssouci, the palace of Frederick the Great. Finally, toward the end of the 1960s, it landed in East Berlin's Tierpark Zoo, where its stands to this day.[137]

"In fifty or sixty years," Goering told a military psychiatrist in Nuremberg, shortly before his suicide in 1946, "there will be Hermann Goering statues all over Germany. Little statues, maybe, but one in every German home."[138]

He would be mistaken. Instead, parts of his Reichsjagdgesetz survived him. To this day, a hunting code of ethics is embedded in the Bundesjagdgesetz (Federal Hunting Act); to this day, the very idea of such a code of ethics is supported by legislation redrafted in 1952; to this day, German hunters follow the German hunting customs that were first set forth during the Third Reich.[139] Goering's legacy is like the sculpture of Raufbold. The golden sheen of old may have disappeared. Yet even hidden, it remains, in the shadows.

6

Not Really Stroganoff

The unknown hero has never been a horse. Even if by fighting and dying
for another's cause, the horse would be entitled to heroism.

—Varujan Vosganian, *The Book of Whispers*

They trekked on by the thousands, duns and chestnuts, blacks and
dapple-grays. Among them were Trakehner warmbloods, the tallest,
noblest, and quickest of the German thoroughbreds. They braved
the summer heat of southern France as well as the winter storms of Russia.[1]
Behind them, as if lost in thought, plodded the brawny coldbloods. These
gentle giants would be the first to be carried off by hunger and disease. At
the end of the caravan came the small, compact Haflinger, valued for their
surefootedness and stamina.[2]

They might have come from a stud farm in East Prussia, where they were
bred specifically for the war. As three- and four-year-olds, they might have
been trained at a riding school for remounts in Lower Saxony, where they
would have learned to suppress their flight instinct and to lay down on command, so as to serve their rider as a living shield during combat. Or they
might been taken from a field in the Allgäu region and become forced recruits
a short while ago, from then on pulling a cannon instead of a plow. Whatever their origins, they had all been selected to drag tons of supplies over
thousands of kilometers toward the east, almost to the end of Europe.

Perhaps some of them were already around in September 1939, when
the Wehrmacht attacked and overran Poland within a few weeks, or at the
beginning of June 1940, when the German cavalry swam through the Seine
and entered Paris a week later. Whether in the west or in the east, they were
there on all the front lines. From 1933 up until the start of the war, the
Wehrmacht increased its stock of equines more than tenfold.[3]

The people back in the homeland, of course, had the impression that war
horses were a thing of the past. Since the beginning of war, the *Wochenschau*

newsreels had been deluding German moviegoers with the image of a mechanized Blitzkrieg, one that was constantly moving ahead with tanks and airplanes. The success stories from Poland, Scandinavia, Belgium, and France reinforced the much-vaunted propaganda myth of a highly modern and virtually invincible German army. Without the horses, though, nothing moved forward, especially on the eastern front. There, as it had been for centuries, the horse was still the real muscle power in the war. Motorized vehicles were not sufficient when it came to getting weapons and provisions as reliably and rapidly to the front.

In September 1939, for example, at the beginning of the campaign in Poland, horses were taken out of domestic agricultural production to meet the enormous demand, whether or not they were needed for the harvest.[4] As with the guard and tracking dogs in the Wehrmacht and the SS, the horses recruited were also given physical fitness exams by veterinarians. If a horse was older than four, stood more than 1.35 meters tall at the withers, was neither totally blind nor lame, and did not suffer from any equine contagion, it was considered fit and then conscripted.

German veterinarians had committed themselves to Nazi ideology and its contradictory notions of *Tierschutz* (animal welfare) soon after the seizure of power and, as a result, toed the official line when it came to recruiting animals.[5] Still, there were a few among them who tried to save as many horses as possible from deployment on the front lines—like the Berlin countess Maria von Maltzan, for instance, a veterinarian, resistance fighter, and "righteous among the nations." She had originally come from Silesia, where as a child she had learned a trick from itinerant Sinti families, one she would now always use before the horses were physically examined: if she stuck a horse with a needle in the leg right above its hoof, then it would immediately go lame. As soon as she pulled the needle out again, though, the paralysis would subside, without having caused any further damage.[6]

By the late 1930s, just under four million equines were living in Germany, which was not enough to meet military demand without fundamentally harming the agricultural sector.[7] Since recruitment was proving increasingly difficult as the war went on, soon more and more horses in the occupied areas were "ausgehoben" ("levied"), as the requisitioning was officially called. In total, from September 1939 to May 1945, almost three million horses, donkeys, and mules would go to battle for Germany.[8]

Crossing the Bug

One such conscripted animal was Siegfried, an approximately seven-year-old, chestnut-colored Trakehner.[9] From far away, he was easily recognizable. He had an unmistakable white blaze, the shape of which recalled a down feather, on his forehead. At a withers height, or floor measure, of 1.70 meters, he towered over most of the other horses quite a bit.

Trakehner horses seemed to be made for the war. Originally, they had been bred as riding horses for the courier service between Königsberg (now Kaliningrad) and Berlin. It took them a whole day less than other breeds to travel the almost six-hundred-kilometer-long stretch.[10] Before World War I, the Trakehner had been the classic horse of the light cavalry. Since the military was the horse's largest customer, for a long time it was able to impose its breeding requirements for a light, quick riding horse, even though East Prussian farmers would have preferred a working animal with a stronger build and a calmer temperament. After the wartime defeat of 1918, when need from the military sharply decreased, farmers regained control over the breeding of work animals.[11] Whereas the Wehrmacht deployed the specimens of heavy stock primarily as draft horses in the artillery, the lighter stocks continued to serve as riding animals.[12] And Siegfried was that kind of beast.

In April 1941, his rider Max Kuhnert, a cavalry officer from Dresden, had gotten his marching orders. Kuhnert was a scout in the cavalcade of the 432nd Infantry Regiment. From the hills of Lower Saxony, they first headed off in the direction of Warsaw. For several weeks the regiment stayed in occupied Poland until they were finally ordered to the banks of the Western Bug on June 20, 1941.[13]

The river constituted the border between areas occupied by Germany and the Soviet Union. In the nonaggression pact that Adolf Hitler and Joseph Stalin had concluded in August 1939, they had agreed to this border and nothing else. Together—with Hitler's troops from the west and Stalin's from the east—they subsequently attacked Poland and divided it among themselves. Most of the soldiers from Kuhnert's regiment refused to accept that they would have to cross the Bug until the last moment, when, on June 22, 1941, Hitler issued the command for the military to begin the long-planned Unternehmen Barbarossa (Operation Barbarossa), the invasion of the Soviet Union designed to create the homeland in the east. Over the course of many months, three million soldiers had marched from the coast of the Baltic Sea to the slopes of the Carpathian Mountains. They were equipped with 1,800 airplanes, 3,600 panzer tanks, 7,000 cannons, 600,000 motor vehicles as well as 750,000 horses.[14]

The men of the 432nd Infantry numbered among the first who had to cross the Bug. Whereas the majority of his unit was to cross over on a quickly constructed pontoon bridge, Kuhnert was tasked with bringing not only Siegfried but also Albert—a ten-year-old, chestnut-brown Hannoverian warmblood that Kuhnert's superior, Colonel Rudolf von Tschudi, rode—safely to the other shore. They could not cross on the bridge, however, for the rocking might have sent the horses into a panic. Moreover, they had been under constant fire. For Kuhnert, therefore, there was nothing left to do but swim across with both horses.

Kuhnert had learned to cross rivers on horseback during his military training, albeit not with Siegfried and not ever with two horses at once. The Bug is a rapidly flowing river. For a while they rode along its shore, until Kuhnert finally discovered a spot where the embankment was not too steep and the river was only around 100 to 150 meters across. After remotely scanning the opposite shore for enemies, he undressed, stowed all his things inside two tarpaulins, put his carbine rifle and his helmet on top so they would not get wet, and roped everything up on the saddles of both horses. Looping Albert's reins around his left arm, he held firmly onto Siegfried's mane and pulled himself up. Sitting on Siegfried, Kuhnert led both horses slowly down the embankment. Albert tried to turn around right away, but once Kuhnert had calmed him down and Albert noticed that even Siegfried was venturing into the water, then he gave in. They moved forward carefully, further and further into the river. Only every so often would one of horses give a nervous snort.

Horses are natural swimmers. Once they could no longer feel the riverbed under their hooves, they began to paddle at a fast trot with their front legs. Then Kuhnert let himself slip from Siegfried's back into the water, so as to swim between the two of them. Doing so, he had to constantly be on his guard not to get struck by the sharp edges of their hooves. He had his left hand on Albert's neck and his right on Siegfried's. Kuhnert thought to himself that they were completely vulnerable the way they were swimming beside one another, an all-too-easy target. What if the Russians noticed them? Worse yet, what if the embankment on the other shore was too steep for them to climb out?

Only Siegfried's and Albert's heads stuck up out of the current. They raised their nostrils high and bared their teeth, as though they could spare only a scornful grimace for the dangerous situation. Yet that is a personification, for horses act that way instinctively so that no water gets into their noses.[15]

Then, as they gradually approached the center of the river, a strong undertow suddenly took them. Kuhnert had underestimated the speed of the Bug's current. He grabbed more tightly onto the reigns of both horses, so that they would not get separated. They were driven several kilometers downstream. Not until the Bug made a wide turn and gradually slowed down were they able to swim the meters remaining to the other shore. They had finally made it. Kuhnert tied up the horses at the edge of a thicket, got dressed, and then set out to find his unit.

Kuhnert and Siegfried subsequently traveled even further eastward, almost to the gates of Moscow. Even though the highways were passable initially and there was forward progress, the effort was nevertheless immense. It took a cavalry unit fourteen to sixteen hours, from sunrise at four in the morning until ten o'clock at night, to cover ninety kilometers. For the most part, breaks were spent watering, feeding, saddling, and unsaddling. In the evenings, the riders had to brush their animals' coats, treat the pressure sores that formed from the poorly fitting saddles, and give them feed and water. And only once they had taken care of all those tasks were they then able to take care of themselves and rest.[16]

The horse played a special role in National Socialist ideology. The horse was the only plant eater, the only flight animal that was venerated alongside the predators they also valued—the wolf that they liked to compare themselves to and the big cats they named their tanks after. Their admiration for the horse can be seen in their art, which was rich in horse motifs from long-legged riding animals that bore their soldiers against all odds to brawny old field nags that stoically plowed the home turf. In Nazi iconography, the horse embodied, in addition to strength, a willingness to sacrifice without appearing submissive in doing so.[17]

Adolf Hitler's relationship with horses was ambiguous. On the one hand, he had one of his favorite sculptors, Josef Thorak, specifically create two larger-than-life horse statues in the classical style. They stood in the garden of the New Reich Chancellery in Berlin, right in front of Hitler's study. On the other hand, however, although he surrounded himself privately with all manner of horse artwork, he avoided any direct contact with living specimens—he downright detested them and considered them dumb.[18] Like "the Russian" who supposedly only worked when under iron-fisted leadership—as Hitler bloviated one night in his headquarters—so too the horse would cast off "in the wink of an eye the rudiments of training" were it not constantly reined in.[19] For him, the horse was impetuous, unpredictable, and, above all, antiquated. The car fanatic Hitler did not think much of the cavalry, either.[20]

The cavalry, however, had a long tradition in Germany. Until 1936, veteran military horses from World War I were awarded a brass plaque on which "comrade in battle" was written.[21] In military circles the cavalry still enjoyed an outstanding reputation even in the 1930s, as it embodied a "chivalric ideal of the warrior on horseback."[22] The connection between rider and horse in war is certainly the strongest manifestation of what the cultural studies scholar Ulrich Raulff calls the "centaurian pact."[23] For millennia there was supposedly no more impressive sight on the battlefields than that of the riding warrior who merged with his war horse into an even more powerful hybrid creature. Even in the more recent tank era, the horse was still considered to be the "icon of what is soldierly," since it embodies obedience and loyalty in addition to its willingness to sacrifice.[24] For instance, General Friedrich Paulus, who became the supreme commander of the 6th Army in Stalingrad in 1942, liked to be photographed for his family album riding high on his horse because he felt the image of a man on horseback befitted the rank of an officer.[25]

Yet the classic equestrian cavalry gradually outlived its service. Only four of the twenty-five supreme commanders serving in the east came from the cavalry. And by November 1941, the last cavalry division had replaced their horses with tanks. The animals did not return home, however. From then on, instead of carrying soldiers into battle, they would pull combat wagons weighing tons.[26] Admittedly, at the beginning of 1943, under the leadership of Georg von Boeselager, a troop of cavalrymen was again set up to fight partisans near Smolensk.[27] Nevertheless, the military role of the horse ultimately shifted during World War II. It would only remain indispensable as a beast of burden, as a draft horse and as a riding animal for messengers and scouts.[28] For this purpose, even the Wehrmacht was dependent on horses in its Moscow field campaign, just like Napoleon's troops had been some 130 years before. To just what extent would become obvious soon enough.[29]

Bogged Down and Frozen Up

Above the Eastern European plain, the late summer sun still had some energy, the ground under Siegfried's hooves was still solid, and the combat wagons beside him still lurched their way forward toward the Russian capital. Yet the first clouds were gathering on the horizon, and at the beginning of September 1941, it started to rain. The rain was so heavy and persistent that the puddles no longer dried up but instead grew wider and wider while the ground became soaked meters deep. A slew of motorcycles and motor vehicles got stuck in the muck. The advance was stalled. Only the horses went on unfazed, pulling the carts from the dreck, though they themselves quickly sank up to their bellies in the morass. After the mire came the cold, a cold that made rivers freeze, engines fail, and, in the blink of an eye, turned Siegfried's warm breath into ice crystals in the air.

At the beginning of October 1941, a flyer from Hitler had nonetheless informed the soldiers on the eastern front that the *Endsieg* (final victory) was within their grasp and that they would reach Moscow in a month. Now, though, almost four months later, Hitler sat in Wolfsschanze, his headquarters hidden in the East Prussian forests, and had to accept that the battle for Moscow was lost. Mud and ice had taken too much out of the German troops. The supply lines had gotten too long, and the few Russian trains they had captured were not enough to compensate for this setback. Because the gauge of the Russian rail network was only just nine centimeters wider than that of Central European tracks, the German soldiers tried to respike the rails to match their gauge. Yet at temperatures of minus forty degrees Celsius, the metal became brittle and broke. In the bitter cold, the machine guns, too, gave up the ghost by the score. There was only one thing they could rely on now, as Hitler grumpily observed in mid-January 1942: "In such temperatures, we're obliged to have recourse to traction by animals," by which he specifically referred to *panje* horses.[30]

The *panje* horse was the stereotypical Eastern European farm horse, and it was something like the Volkswagen of the eastern front. "Panje" in Polish has the approximate meaning of "master" and was soldier's slang for Polish and Russian peasants. Standing not even 1.5 meters high at its withers, the small tenacious horse was barely larger than a pony, yet it could cover up to 150 kilometers a day without getting worn out. On top of that, it was not prone to falling sick, despite temperatures of minus fifty degrees Celsius, and was content with barley and corn as substitute feed. If necessary, it would even make do with the tree bark that it gnawed off the trunks used for log cabins.[31] Before the Russian campaign began, every German division had been assigned two hundred to three hundred wagons with *panje* horses to increase their mobility. Soon enough they constituted 20 percent of the entire horse stock in the *Ostheer* (army of the east). The *panje* horse's only disadvantage was its build. On account of its short legs, it could only run at most ten kilometers per hour in a trot. In the same amount of time, a horse like Siegfried would cover up to fifteen kilometers. On top of that, the *panje* horse could carry at best loads weighing 75 kilos and pull at most 150 kilos, and so it was simply too weak to pull the Wehrmacht's most commonly used cannon, the light field howitzer.[32]

For Max Kuhnert and his horse Siegfried, the goal was no longer Moscow. Their unit was already on the retreat and had spent several days riding southwest of the Russian capital, somewhere in the western Russian wasteland. In the deep of winter, it became even more costly for Kuhnert to come by enough feed for his horse. Siegfried needed around ten kilos of hay and oats every day in order to keep up a minimum of strength. Trakehner, in fact, lose weight quickly once they are no longer getting enough to eat.[33] On some parts of the front, the draft horses even had to be supplied with hay and oats by airplane.[34] Many of them were already so severely emaciated from long marches and poor nutrition that the saddles and harnesses no longer fit them properly and their backs had become abraded.[35] Although Siegfried had slimmed down as well, until then Kuhnert had still managed to get him through, even if he came close to despair on the last night of 1941. They had pitched camp in the vicinity of the village of Yukhnov, in the Kaluga Oblast. For hours he had been looking for a serving of oats or something similar but found nothing. The only thing he could scare up was an old sofa. He promptly ripped it open and picked out an armful of dusty straw that had served as padding, and Siegfried willingly ate it up.

Time and again, besides procuring feed, another big challenge every day consisted of finding dry shelter for Siegfried for the night—not exactly a simple proposition for a horse of his size. Most stalls were just high enough for the local *panje* horses. One evening, while yet another snowstorm was raging around them and little icicles had already begun to form on Siegfried's eyelashes and nostrils, Kuhnert spied a crooked shed at the edge of a field. In front there was a small counter window, and a narrow door in the rear. Presumably fruit and vegetables were sold there during the summer. Now it stood empty. While it continued snowing outside, Kuhnert tried to coax Siegfried to move little by little into the squat little shack. In the end, he would just fit into it diagonally.[36]

Even for horses, the Russian field campaign had become such an ordeal by the middle of March 1942 that a majority of them barely managed more than ten kilometers per day.[37] Many lost their shoes along the way, but there were not enough farriers and material to put new ones on them.[38] The animals broke down by the score, loaded down by heavy steel combat wagons, because on the unpaved streets the caissons' wheels would dig deeply into the ground that had softened from the thaw. As a result, the soldiers had taken to calling the combat wagons "Pferdemörder" ("horse murderers").[39]

Many horses died of colic, heart disease, or strangles, a bacterial infection of the upper respiratory tract. Others had their entire coat so devoured by scabies mites at times that it was hard to tell whether the afflicted animals were blacks or dapple-grays.[40] In the swamplands of Russia and the Balkans, moreover, a disease appeared that was not widely known in Germany, one that spared the indigenous *panje* horses, admittedly, but affected the sensitive coldbloods in particular: piroplasmosis, also known as "horse malaria," was carried by ticks and was the deadliest horse contagion in a few regions of Russia.[41]

In order to care for the large number of horses in the military, veterinarians were increasingly ordered to the front. By 1939, the caravan of the veterinary service comprised fifty-one thousand people, climbing in the years that followed to just short of ninety thousand. In total, almost an eighth of a million veterinarians and farriers had been drafted.[42] The need for specialized personnel was so great that over the course of the war, more than eight thousand of the approximately ten thousand veterinarians living in Germany were drafted for a certain period of time. In addition, their lives were at great risk, for unlike medical practitioners and nurses for people, veterinary practitioners were not considered to be specially protected personnel according to the Geneva Convention. On top of that, veterinarians wore crimson red epaulettes on their uniforms, making them easy to confuse with members of the general staff. Approximately one out of every six of them would not return home.[43]

In the spring of 1942, Max Kuhnert and Siegfried were still with their unit stationed near Yukhnov in western Russia and doing well. Until now,

Siegfried had gotten off lightly, except for a bad case of colic on account of lack of water and a wound he suffered on his right rear hock when he ran into a panzer tank. Because disinfectant was hard to come by, Kuhnert first cleaned the wound with water and then peed on it.[44]

In cases of grave injuries, the military veterinarians resorted to even rougher methods. Because there were hardly any anesthetics, the so-called *Nasenbremse* (twitch; literally, "nose brake") was employed with injured animals. It was a simple wooden handle with a rope loop that was laid across the horse's upper lip. The loop pulled so tight when the handle was twisted that the pain it caused displaced the pain of the operation and thus put the horse at ease. At least in theory. In reality, however, the stress for the horse was immense.[45]

Still, there was a war going on, and since soldiers were losing their lives day after day, the consideration given to horses was that much less. Max Kuhnert and Siegfried would soon experience that in a bitter way. It happened one morning. He had already been saddled, and Kuhnert was just about to put in the bit, the mouth part of the bridle, when the roll of thunder resounded in the distance. In the very next instant, a mortar shell landed next to them. Kuhnert was yanked to the ground by the impact of the explosion. Once he made it to his feet again, he looked down at himself—only his coat had been torn and burnt. Then he saw Siegfried. The Trakehner was standing there quietly. Yet under his right eye, right where Kuhnert's hand had been resting, there now was a gaping wound. Siegfried turned his head toward the left saddle bag, where blood was already streaming. Very slowly his front legs gave way, and then his powerful body fell on its side. With bulging eyes opened wide, Siegfried stared at his rider, as if, so it seemed to Kuhnert, wanting to say farewell.

Kuhnert did not want to believe it was true. Beating on Siegfried's neck, he screamed, "Get up! You can't do this to me!" Yet Siegfried's eyes had long gone vacant. Crying, Kuhnert fell to his knees next to him. He stroked Siegfried's mane, sensed how the warmth was gradually leaving his body. All those months that they had spent together, which had felt like a whole lifetime to him, Siegfried had been Kuhnert's protector, his comrade. How many times had Siegfried saved him by turning his ears a certain direction or suddenly snorting whenever he noticed something unusual? "He wasn't just a horse to me," Kuhnert would write in his memoirs. "He was my best friend." Kuhnert was not paying attention anymore to the shells that kept landing around him; he just kept petting Siegfried's ears. Later, when a few soldiers came by with another horse to take the lifeless bodies away, Kuhnert still could not comprehend it. "Be careful," he called to them, as if Siegfried had only been lightly wounded.[46]

The fate of the horses mattered to a lot of soldiers, even those who were not as closely connected to them as Kuhnert had been. In countless letters from the front, they painted the suffering and dying of the war horses, as did the student Harald Henry from Berlin, who wrote: "Torn apart by grenades, bloated, their eyes having rolled out of empty red sockets, standing and trembling, leaking slowly from a small hole in the chest, yet unstoppably bleeding—that's how we've seen them now for months. It's almost worse than the human faces that have been ripped away, the burnt-up, half-charred corpses with their bloody ribcages split open, worse than the narrow streaks of blood behind the ears or on the face of the mutilated."[47] In the Wehrmacht, a poem to the soldiers' *Kamerad Pferd* (comrade horse) that lionized the animal's selfless sacrifice made the rounds. It went:

> You'd eat rotting thatch from the roof for your feed
> And starve even more, my dear trusty steed,
> Burnt up by the fire, and wounded you'd bleed
> But we'll keep you in our hearts, our dear trusty steed.[48]

Lamenting the horses not only served the sacrificial cult surrounding the suffering, innocent creature, as the cultural historian David de Kleijn has written, but also referred "back to those who lamented." In pitying the horses, then, perhaps the soldiers were expressing self-pity and the silent sorrow for the pain they themselves suffered, too.[49]

Meat Stew and Paprika

Not all soldiers had such a deep relationship with animals, and in view of their own suffering, it must have been difficult for many combatants to feel any compassion for the beasts. The longer the war persisted, the more frequently those comrades-in-arms would mean the undoing of their trusty steed "Kamerad Pferd." In no place was that downfall as evident as in the town that comes second to none in symbolizing German defeat.

After the battle for Moscow was lost, the Wehrmacht attempted to regain ground with a 1942 summer offensive. Yet the 6th Army that was to capture the industrial city of Stalingrad, given its strategically advantageous position on the Volga River, was hopelessly undersupplied. In November 1942, the Red Army had succeeded in bottling it up tight. As a result, a quarter million German soldiers were entrapped—and with them fifty-two thousand horses and mules.

Ever since the fall of 1941, the Wehrmacht had been feeding its prisoners of war with horse carcasses.[50] Yet the more desperate the German army's situation in Stalingrad became, the more these animals had moseyed into their own stewpots. Just a few days before Christmas 1942, full of hope, one soldier wrote to his parents, "As long as we still have horses, things are okay and, besides, the führer won't leave us behind."[51]

Even so, Hitler demanded that the soldiers stick it out. They were to break through the entrapment by themselves, an undertaking that would, however, fail over the course of that December. To keep the soldiers alive, four thousand horses from the allied Romanian cavalry were slaughtered on the spot and made into stew. Additionally, every man got two slices of bread per day.[52] The starving soldiers sought their refuge in sarcasm, christening the stew "Horst-Wessel-Suppe" ("Horst Wessel soup")—for just like the comrades in the eponymous SA battle hymn, the chunks of meat in the thin broth only marched "along in spirit."[53]

Once the mercury in the thermometer sank to minus fifty degrees Celsius, the strength of both man and beast noticeably dissipated. "The last horse was eaten up long ago and no idea if this shit come [sic] to an end," one soldier wrote in the middle of January 1943.[54] To keep from starving to death, the soldiers then wolfed down even those animals that had been lying around dead for weeks and were gradually starting to decompose.[55] In order to dismember the carcasses, they set off hand grenades in their bellies and then boiled the blown-out scraps of flesh in water melted from snow.[56]

Two weeks later, in February 1943, when the half-frozen, famished 6th Army laid down its arms in the rubble of Stalingrad, only a little more than a quarter of the 250,000 German soldiers that had been ensnared were still alive. Only six thousand of them would later return home from captivity as prisoners of war.[57] Not one of their fifty-two thousand horses made it back from the cauldron that was Stalingrad. All of them froze to death, died on the battlefield, or were devoured.[58] Even in places where the fighting had long been concluded, horses continued to die. Strictly speaking, it would only reach its sad climax in May 1944, in Crimea.

Two years prior, in July 1942, the Wehrmacht had captured the Black Sea peninsula after heavy fighting. In Hitler's delusions of a Germanic *Weltreich* (world empire), Crimea occupied a special role. He planned, after beating Stalin, to settle people from South Tyrol (in present-day Italy) there, so as to reinstate the *germanische Tradition* (Germanic tradition), which had supposedly begun in the region with the Crimean Goth *Volksstamm* (tribe) who lived there in the third century. Going forward, Crimea was to be called Gau Gotenland (Gothenland District) and the town of Sebastopol was to bear the name Theoderichshafen, in memory of Theoderich, the legendary king of the Goths.[59]

The idea came from Alfred Frauenfeld—the *Gauleiter* (district leader) of Vienna, designated to become *Generalkommissar* (general commissioner) of Crimea—and was promoted by Himmler, who informed Hitler of it. The resettlement was to be initiated after the war ended.[60] Specifically for that purpose, Himmler had begun a research program in the SS Forschungsgemeinschaft (Association for Scientific Research) called "Deutsches Ahnenerbe" ("German ancestral heritage) in order to breed winter-hardy steppe horses for these settlers in the east.[61] In the summer of 1943, Himmler tasked the SS horse expert Ernst Schäfer with starting a breeding program for this new *Pferderasse* (literally, "horse race").[62] Rather quickly, however, it became obvious that nothing would come of any of it. South Tyroleans in Crimea remained as much a flight of fancy as did a new breed of horse.

Ever since the fall of 1943, the soldiers of the 17th Army stationed in Crimea had been primarily involved with defending it against reinforcements from the Red Army. They had been forced to yield one position after the other, and by that point, they were only just able to hang on to Sebastopol. There in the strongly fortified harbor town on the southwestern coast of the Crimean peninsula, they held out until the beginning of May 1944, when the situation finally became so desperate that they received the long-desired command to evacuate.[63] Now, however, the question arose as to what would happen to the thirty thousand horses that were with them.

Among the horses in Sebastopol was Paprika, a dapple-gray mare. She came originally from the town of Barlad in eastern Romania. Her rider had discovered her among some indigenous riding animals there in June 1941, which the Wehrmacht had purchased to cover its increased demand and to compensate for its own losses over the preceding months. Neither the name nor the rank of the soldier is known. We only know his memories of Paprika that he wrote down.

"Paprika, do you still remember?" he wrote in a letter about their shared experiences. "We understood each other from that very first day. You were clever, you had spirit. You responded to the slightest leg pressure. Of course, your trot was terrible. Why did you have to raise your front legs so high? Don't be angry with me, Paprika, I often suspected you came from the frivolous circus world. But your splendid, incomparable gallop, your speed, your jumping ability were world class."[64] Nobody had dared to ride the impetuous animal, except for him. She bit, and she bucked as soon as any other horse stood next to her. Nevertheless, he kept her and with time won her trust.

Paprika accompanied him on the more than thousand-kilometer march from Romania, across Bessarabia (in present-day Moldavia) and Ukraine, all the way to the southwesternmost tip of Crimea. During every longer break, she lay down next to him, and he laid his head on her belly. Then his comrades would taunt, "That beast will kick your bones to bits one day," but he only grinned back scornfully. What did they know? They did not understand, either, when he and Paprika performed a little trick whereby she carefully took a morsel of bread from his mouth with her teeth. The others only shook their heads and said, "One day that beast will bite your nose off."

Nothing like that happened. Instead, he received the Iron Cross for the successful post rides he made on her back. When they landed in a minefield on the Kerch Peninsula at the eastern edge of Crimea, he had her to thank, above all, for her calm and precision in helping them get out in one piece. Because the barrier tape used to designate landmine areas had been ripped away, they had had to move backward. "With your ears perked and softly snorting, you backed out, absolutely slowly, in your own tracks," he wrote. "I never said to you, Paprika, that after a happy end to the war I wanted to buy you, that I had already found shelter for you with good people in Berlin. Now we have to part, the destiny of an unrelenting war is tearing us apart."

It was his farewell to her. Though Paprika so often may have saved his life, her own would end here. The 17th Army's command to retreat was a death sentence for her and the other horses. As per the command, they were not to fall into Russian hands. Because the soldiers could not take them along in their escape, they were to be "liquidated," all thirty thousand of them. Many of the riders, however, refused to shoot their horses to death personally. The men of the veterinary company undertook the task instead. They lined up the horses individually next to one another on the cliffs of Severnaya Bay. "Once more I inhaled Paprika's warm breath," her rider wrote, "once more I laid my face on her velvety-soft nostrils. I watched her go until the seawall blocked my view."

As close as the relationship with their equine comrades-in-arms may have been—the soldiers' zombie-like obedience outweighed it. One horse after the other was taken by the bridle, then they put the gun barrel to its ear, pulled the trigger, and subsequently pushed it over the edge of the cliff into the sea. In the end, because the executions were dragging on and the horses were getting more and more restless, the remaining animals were all herded together and riddled with machine guns, until not one was left standing. Long afterward, their carcasses would still be floating around in the bay, where the sea swells dashed them over and over again against the cliffs.

On average, every day the war was raging, 865 German military horses lost their lives.[65] At the end of the war, the total would be 1.8 million. Three-quarters of them died in battle, like Siegfried.[66] To be sure, the Trakehner warmbloods proved to be particularly tough; indeed, by the fifth year of the war some of them had served for more than twenty years.[67] Nevertheless, at around eight years old, Siegfried had clearly lived longer than most of his kind. For a horse on the German side of war rode out only four years on average, before bullets, disease, or the cold carried them off. In that regard, they still lasted longer than most engines; on account of wear and tear, motor vehicles gave up the ghost after one year on average, holding up for not even two months toward the end.[68] Six decades later, the historian Reinhart Koselleck, who as a soldier in the Wehrmacht lived through not just thousands of horse deaths, would sum up the dilemma of this war in the following words: "It could not be won with horses, and even less so without them."[69]

Germany—Horse Country

There it stands, alone and without any saddle, not a rider in sight. Its flared tail swishes as it looks a touch too proudly across the courtyard. Surrounded by stone walls and pieces of rubble, its faraway gaze remains suspended, fixed on the wall of the building opposite it. Bronze and larger than life, this horse stands in Munich's Schönfeldstraße, the street that flanks the parade court of the erstwhile Bavarian War Ministry, on a basalt pedestal.[70] The inscription on the back of the pedestal reads, "Der deutschen Kavallerie zum Gedenken" ("In commemoration of the German cavalry"), and the front is engraved with the dates "1870–1945."

The memorial was created by the sculptor Bernhard Bleeker and commissioned by former cavalry associations.[71] The 100,000 Deutschmarks to erect it were collected by the Verein zur Errichtung eines Denkmals für die Gefallenen der Kavallerie-Regimenter (Society for Erecting a Memorial for Those Fallen Members of the Cavalry Regiments)—specifically founded for this purpose—as well as the Free State of Bavaria, as the province is officially known.[72] Horses are an essential motif in Bleeker's work.[73] He produced many designs for the statue and reworked them over and over again. At one point, the horse had a saddle, another time a steel Wehrmacht helmet lay beside it, and in yet another version, carrying handles were attached to the pedestal so that it looked as if the horse were standing on a gigantic coffin.[74] Later on, Bleeker would maintain that he had supposedly wanted to create a memorial for the "unbekannte Kavallerie-Pferd" ("unknown cavalry horse"), which would stand for the many horses he had seen come to their end during World War I. At bottom, however, it is about those who cannot be seen. And not about horses.

The fact that the sculpture turned out to be so modest in the end and did not have any military attributes was not a coincidence. In the relatively young Federal Republic, any pathos regarding war would have been out of place. Nevertheless, on May 29, 1960, the dedication day, a shadow still fell over the square. Bleeker, the creator of the statue, had not only joined the Nazi Party in 1932 but had, on top of that, been on the *Gottbegnadeten-Liste* (list of those graced by God)—the index in which Joseph Goebbels and Adolf Hitler had registered artists whom the regime, as late as 1944, wanted to be exempted from wartime service on account of their especially valued work.[75] Now, fifteen years after the war ended, Bleeker had long been classified as just a *Mitläufer* (a nominal Nazi fellow traveler) and had been reinstated as a member of Munich's Akademie der Schönen Künste (Academy of Fine Arts). The dedication speech for the horse sculpture was delivered by Dietrich von Saucken, a former Wehrmacht general, who had only returned from captivity in 1955, one of the last ten thousand prisoners of war in the Soviet Union. In his tribute, he recalled the "German soldier horses" that had served "devoutly, willingly, and persistently until their last breath," and highlighted the cavalry's "spiritual substance."[76] What von Saucken certainly must also have been referring to was all the other soldiers in the Wehrmacht. In creating a memorial to commemorate all German cavalry soldiers since the 1870–71 war between Germany and France, they were able to memorialize their fallen Wehrmacht comrades at the same time, too. The Pferdedenkmal (Horse Memorial) is a fig leaf covering all the old school die-hards.

The horse, therefore, did not quite entirely escape its role as a symbol of war; indeed, the memorial embodies what the writer Elias Canetti imagines as the ambivalent relationship between human beings and horses: "The finest statue of man would be a horse that has thrown him off."[77]

Except for Bleeker's bronze horse and a very few other monuments, not too much has remained to recall the war horses of yore. Their descendants, however, still serve in the military even today. Located in Bad Reichenhall, in Bavaria, is the Bundeswehr's Einsatz- und Ausbildungszentrum für Tragtierwesen 230 (Operation and Training Center for Pack Animals 230), which grooms pack and riding animals for war. Where the terrain is too impassable for vehicles or the climate too extreme, mostly mules—able to haul loads of up to 160 kilos even at heights above five thousand meters—are deployed. With the help of these pack mules, the Bundeswehr supplied the sentries in Kosovo along the border of North Macedonia, from 2002 to 2004. In the Afghan province of Badakshan in 2009, moreover, they tested the deployment of donkeys as pack animals in high mountain ranges. As riding mounts for reconnaissance, they continue to rely on tenacious Haflinger horses, just as in World War II.[78]

Nevertheless, the horse leads only a niche existence now. It and its relatives have disappeared from the battlefields by and large. The horse left our everyday lives long ago for the equestrian sports scene and its magazines, where it stands as a costly hobby. The only commonplace reference that has remained is an abbreviation for an engine's performance: HP—two letters to recall horsepower, the erstwhile muscle of the war.

Meanwhile, how emotionally connected we still are to the horse was demonstrated in the spring of 2013. In supermarkets in multiple European countries at that time, purported beef products turned up that in fact consisted to a great extent of horse meat. An outcry went throughout Germany.

A headline in the tabloid *Bild* on February 14, 2013, read "Pferde-Fleisch in unserem Essen!" (Horse meat in our food!) but was at the same time accompanied by a subtitle asking "Was ist an Pferde Fleisch so gefährlich?" (What's so dangerous about horse meat?) Even in the more serious media outlets, the fact that the affected pasta sauces and frozen lasagnas also were partly contaminated with phenylbutazone, a horse-racing drug, was more of a side issue.[79] One week later, the German weekly *Die Zeit* stated, "The lack of any declaration, that is, the betrayal of the consumer, is the actual scandal."[80]

The reasons for the bad reputation of horse meat, even today, are complex. On the one hand, it was once considered poor people's food, and its salubriousness was often doubted. On the other hand, the horse had never served purely as a source of meat, since its high need for feed was simply too expensive.[81] To be sure, at the end of the 1950s in Germany, around twenty thousand tons of horse meat were produced. Yet as the postwar years grew distant and Germany became more prosperous, the traditional rejection of horse meat became even more intense.[82] Behind the rage of the 2013 affair, however, was historical experience and not just any "little girl's hysteria," as the daily paper *Die Welt* observed, writing: "Nowhere is the disgust for horse meant so pronounced as in the generation that lived through the Second World War and the years that followed."[83] Similarly, cultural historian Peter Peter has perceived in this aversion "an unpleasant memory, in the case of older people in particular," of "the emergency slaughtering of horses, of images of Stalingrad."[84] These memories have presumably entered into the collective memory as well, like recollections of the nightly bombings and the hunger winter.

The especially emotional reaction to the horse meat scandal may also stem from the fact that it not only conjured up people's memories of their own victimhood but also brought their own dark sides into light. For, according to historian Rainer Pöppinghege, the horse symbolically stands for those victims of the war "who, out of shame, were not mourned."[85] As a result of an unexpected trigger, then, the long-repressed horror of that time when many in need did not spare their animal companions—and some not even their human comrades—was brought to the surface.

Epilogue

Until the Last Dog Is Hung

When I went for water a house fell on me
We bore the house
The forgotten dog and I.
Don't ask me how
I don't remember
Ask the dog how.

—Inge Müller, "Under the Rubble III"

The end. The Red Army had crossed the city limits. Only a few days remained, if not mere hours, until Berlin would fall. The walls of the Reich Chancellery shook with increasing frequency from the impact of the blasts. Underneath, protected by meters-thick reinforced concrete, Adolf Hitler was making final arrangements.

"One of these days I'll only have two friends left," he had often said to Albert Speer over the past two years. "Fräulein Braun and my dog." Hitler seemed both disdainful and disappointed at the same time. Speer felt insulted personally, though admittedly he thought that in a certain sense Hitler was right. To be sure, if Hitler was right, it had less to do with Hitler himself than with the "staunchness of his mistress" and the "dependency of his dog."[1]

The day before, on April 29, 1945, Hitler had rewarded Eva Braun, his longtime partner in life, for her "staunchness" and married her. Today, they would together freely take their own lives. First, however, it was Blondi's turn. The German shepherd who had lived with him for three years now had hardly left his side in the past few weeks. Even an egomaniac like Hitler had noticed it. Dogs, he always said to his secretaries, are more faithful than people. In those days, the impression he gave was mostly apathetic: he holed up in his study for hours, listened to Wagner operas, and stared up at the portrait of Frederick the Great hanging above his desk. Hitler venerated the Prussian king, seeing a sort of kindred spirit in him. Time and again he quoted his saying: "The more I see of men, the better I like my dog."[2]

And what had Hitler not done for his dogs? Once, he had summoned the world-renowned surgeon Ferdinand Sauerbruch to Obersalzberg to operate on one of his German shepherds.[3] Another time, when Blondi was at a Munich veterinary clinic with an infectious disease, he had bulletins sent every day to update him on the condition of her health.[4]

In the final months of the war, when the only news reaching him from every front was of retreats and defeats, when he believed himself to have been abandoned and betrayed by his own generals, he hardly spoke of anything else but dogs: of Blondi, above all, and of her planned "wedding," as his secretary Traudl Junge later recalled.[5] Blondi was finally supposed to have offspring. After multiple attempts with various stud dogs, things had finally worked out and, at the beginning of April 1945, she brought five whelps into the world. Hitler immediately took one male puppy for his own and gave it the name Wolf. Often, as his secretary Christa Schroeder recalled after the war, he would sit lost in thought, petting Wolf while whispering his name.[6]

Hitler had been holding out in his bunker since January 1945. He only left it for a few minutes every morning to take a walk with Blondi around the garden of the Reich Chancellery. By this time, the war had long been lost and was just one bloody retreat. People from the eastern areas of the Reich were fleeing the Red Army by the hundreds of thousands. In East Prussia, endless trails of people trekked across the frozen Vistula Lagoon heading for what was then Königsberg (now Kaliningrad), hoping to catch one of the refugee ships there. The refugees stowed all their worldly possessions onto covered carts and sleighs that were drawn by their horses. The ice broke under the weight of all the convoys. People and horses sank into the brackish water, sliding under the ice floes, freezing and drowning to death. Others were gunned down by enemy fighter planes and tanks.

Little would remain of the Trakehner stud farms, once the pride of the region. From a herd once comprising thirty thousand animals, only eight million mares and forty-five stallions reached the west, ending up widely scattered across Germany, along with the refugees.[7] East Prussia, "the classic reservoir for Germany's horses," as the German newsweekly *Der Spiegel* wrote in 1951, would henceforth be "Russian occupied."[8]

From Pomerania, too, people fled en masse at the start of 1945. In his notes, Hans Schlange-Schöningen described the final months of war in the way he had also already described the past few years—with a proper dose of fatalism: "The carts are being loaded. Most of the people are panic-stricken. I shall do my best to see that they go off in the best possible order. My brave wife and I will stay on. Hitler's officials are flying, of course." Soon afterward, endless flocks would cross the Oder River, where "the cold is still intense, and snow-storms are frequent again. . . . Thousands of horses are scattered along the sides of the roads. Dead people have been temporarily buried in the snow. And all the time the flight into the interior goes on. The Russians are coming! Napoleon's retreat from Moscow must have been child's play by comparison."[9] While his estate overflowed with refugees, Schlange-Schöningen attempted to care for the animals as best he could, along with the remaining foreign workers. Soon, though, he also would have to leave his ancestral estate forever. While an SS division was plundering the farm and nearby locality, he escaped with his family to Holstein.[10] There he would write, "Schöningen was a burning heap of ruins. Now I was the fugitive stranger."[11]

He did not write what became of his animals, though in the final chaos of war they would have been all too easy pickings. At that time, in vast parts of Germany, the pig population decreased dramatically.[12] In Saxony, for example, the stock fell from around one million animals at the start of the war to two hundred thousand in 1945.[13] Presumably, what also doomed them was the fact that they were neither suited for long journeys nor as beasts of burden but, rather, for just one single purpose—being slaughtered and eaten.

Thousands of *panje* horses that had accompanied the Red Army to Berlin ended up staying there after the war, where they would characterize the streetscape going forward. In August 1945, the newspaper *Neue Zeit* noted that "in long, immeasurable lines, wagons rattle across the torn-up pavement of the city in ruins" and how "the country came into the city; we have moved much closer to nature. All of our lives, including the economy, are entirely dependent on the horse."[14]

Yet even when the war was over, the suffering of the horses still did not end, as the *Neue Zeit* reported in December 1945: "Their gleaming coats have become shaggy, and protruding ribs and hip bones show only too clearly that their stomachs are often rumbling. . . . Their horseshoes clatter or are even completely missing; ill-fitting harnesses chafe wounds as large as the palm of your hand on the animals' heads, necks, and bellies." Every day they were hitched once more to carts without any consideration for their health. "Perhaps we have gotten too used to the mechanical insensitivity of engines; otherwise many a wagon owner would certainly have a more loving regard for the hard life of his draft horse."[15]

The Total Animal

If there was any animal at all that profited from this war—at least for a certain period of it—then it was definitely the body louse (known in German, more properly speaking, as the clothes louse, i.e., *Kleiderlaus*). Its story could tell the entire history of this total war. For wherever the war raged, it was there too. In their recollections and retrospectives, generals and historians hardly had a word to spare for the little bloodsuckers, yet the letters from the soldiers in the field were teeming with them. According to the Stalingrad veteran Wilhelm Raimund Beyer, anyone who supposedly had nothing to say about lice "was not at Stalingrad!"[16]

The letters written to those at home reveal that initially the lice provided the infantry grunts with proof of those prejudices they had brought along with them, namely, stereotypes about the backward "filthy" Russians as opposed to the cleanliness and orderliness of Germans.[17] With time, however, many among them were forced to acknowledge that the lice did not care whom they bit. And that they nearly drove you insane: "Larvae in your clothes, lice in your laundry, and fleas everywhere in between! No matter how many times you undress and go over it all, afterward it's the same thing," one soldier wrote. Another soldier asked his folks at home for help: "If it's possible to acquire some kind of salve or something similar that you can use as a repellent, please send it." And yet another soldier survived through sarcasm, remarking tersely: "It's going great for our lice; they're multiplying nonstop."[18]

Even when the guns were silent, they kept on biting, incessantly sucking, never resting. The louse became the biting irony of history, its beast made flesh, if you will, the total animal. It defied all ideologies. In that way, friend as well as foe were united in their itching. For "Rassenhygiene" did not help, either, against lice in one's uniform. The louse made no distinctions, whether in the air raid shelters of bombed-out German towns, in the frozen foxholes of Stalingrad, or in the barracks of Auschwitz. Some put it to good use. So, for example, Italian Auschwitz survivor Primo Levi reported how the concentration camp's laundry women would collect the clothes lice from the dead and put them under the collars of the freshly pressed SS uniforms, so as to infect the guards with typhoid fever and typhus. For lice, as Levi wrote, "are not very attractive animals, but they do not have racial prejudices."[19]

Blondi's End

In the bunker at the Reich Chancellery, meanwhile, Hitler called for his acting personal physician, Werner Haase, as well as for Fritz Tornow. Tornow was around forty years old and originally came from Silesia (in present-day Poland). He was average size and weight, with dark blond hair and an oval face. He had a narrow mustache and a set of false upper teeth.[20] Though Tornow was a noncommissioned officer, he in fact occupied a far more significant position in Hitler's entourage. He was his dog handler.

Hitler may have doted on Blondi, but Tornow was the one who took care of her most of the time. At Obersalzberg and at Wolfsschanze in East Prussia, he had spent hours training her. Tornow was especially tasked with going on walks with her during the summer months, when Hitler preferred to stay inside the cool of the masonry walls.[21]

Hitler had given up the fight some time ago, but he still feared that after his death he would be put on display "by the Russians in a panopticon" or as "an exhibit in the Moscow zoo." Therefore, nothing of his corpse was to remain.[22] Nor did he want his dog to fall into enemy hands. The very thought of it made him sick.[23] When he made his exit, his dog was to go with him.

In case of emergency, Hitler had acquired small cyanide ampoules from the SS, but he had his doubts as to whether the poison was dead certain, as it were.

So Hitler asked Werner Haase how they could test whether the capsules worked.

On a dog, Haase answered.

It was already midnight when the dog handler Tornow led Blondi into the lavatory wing of the bunker. Then everything went very quickly: while Tornow held open the dog's mouth, Haase took one of the capsules and crushed it with a pair of pliers in her throat. The smell of bitter almonds rose upward. Blondi started to stagger, then collapsed convulsively. Thirty seconds later, she was no more.[24] It was only then that Hitler entered the room. Without moving, he looked at her lifeless body, not saying a word. After a little while, he turned around and left.

Tornow's task, however, was not yet completed. While Hitler locked himself up in his study, Tornow climbed up the stairs into the garden, where Wolf and the other whelps had been lined up next. The children of Propaganda Minister Joseph Goebbels were not at all pleased that they were to give up the playmates they had passed the time with those last eight days. They had no idea that they would follow them soon enough.

Besides the five whelps, Tornow also shot to death the black Scottish terrier that Eva Braun brought into the bunker with her, as well as the dog of Hitler's secretary Gerda Christian and Tornow's own dachshund. Then he got drunk.[25] Hitler would kill himself and Eva Braun on the afternoon of April 30. When the Red Army captured the Reich Chancellery the following day, Tornow and the other survivors surrendered, without putting up any resistance.[26]

A few days later, a Soviet search party inspected the grounds. In the garden, the soldiers found the charred body of a man and a woman, as well as the remains of two dogs. One was supposedly a German shepherd whelp, the other a fully grown animal. Though the collar was admittedly sooty, the inscription could still be read: "Always with you."[27]

In the end, the self-proclaimed animal lover Hitler revealed the true nature of his relationship with dogs—they were there to obey him and make him feel like they were faithfully devoted to him. And because he, too, no longer saw any meaning in life, in his eyes there was no reason for Blondi to live any longer. Scorched earth, everywhere.

What in fact might Frederick the Great have said about this behavior? In a letter to his sister Wilhelmine in 1752, the Prussian king—who had not only abolished torture but also attributed a spiritual life even to animals—wrote: "I believe a human being who can be indifferent to any faithful animal will not be any more grateful toward his own sort and that, if one is faced with the choice, being too sensitive is better than being too harsh."[28]

The Time of the Wolf

Even though the Third Reich had been destroyed, the German shepherd would never quite lose the reputation associated with it. In the culture of international pop, it is considered to be an inalienable Aryan accessory.

Going forward, it would belong to the image of Nazism, as the historian Wolfgang Wippermann argues, because "the picture of the Third Reich does not seem perfect" without it.[29] It has hardly been any different for its progenitor. Wherever right-wingers and fascists in Germany have gained influence politically since, both they themselves and even their opponents have used the wolf as a symbol. In 1965, for instance, the singer-songwriter and peace activist Franz Josef Degenhardt responded to the rise of the radical right-wing National Democratic Party of Germany with his song "Wölfe mitten im Mai" (Wolves in the middle of May).

In the months immediately following the war's end, the wolf was initially considered to be, above all, the symbol for crude customs in a country that had been laid to waste. At the time, Germans spoke of the *Wolfszeit*, the time of the wolf, during which people only looked after themselves and their families and distrusted strangers, in short, the time "when humans [became] wolves to humans."[30]

It became all too convenient to be able to blame everything abysmal and predatory from those years on a flesh-and-blood wild animal. That is what happened in the Lüneburg Heath in 1948. Since the spring of that year, incidents of poaching had been piling up in the Lichtenmoor bogs between the Weser and Aller Rivers. To the dismay of farmers there, multiple cattle, sheep, and goats met their end with puzzling wounds. Rumors spread like weeds. Besides stray dogs, some claimed to have seen an escaped puma or tiger; in fact, there was even talk of a werewolf. The media soon found a suitably sensational name for the "rätselhafter Ungeheuer" ("puzzling beast")—"der Würger vom Lichtenmoor" (the Lichtenmoor strangler), a moniker that had also appeared a popular horror film from around the same time (i.e., *Würger im Nebel*, released originally in English as *Strangler of the Swamp*, dir. Frank Wisbar [1946]). Yet behind the slashes, the farmers primarily suspected their archenemy—the wolf.

It had happened that every so often, in the years before Germany was divided, lone wolves would wander all the way to western Germany in their search for food. It was questionable, however, whether any wolf was to blame in the incidents of 1948. The wounds of the animals killed were just too clean, as if they had been cut with a scalpel, not ripped open with teeth. A hunt lasting weeks began, the biggest in the history of Lower Saxony. Expressly for this purpose, the hunters, having been disarmed since the end of the war, were given back their shotguns from the occupying British forces so that they could finish off the *Eindringling aus dem Osten* (interloper from the East).

Another, more human cause was obvious. At this point, it would be months before currency reform. The past two so-called hunger winters were still felt deep in the bone of many; meat was being rationed and traded on the black market. It was the time when stealing was just called "Hamstern" ("hoarding," like a hamster does using its cheeks). After June 1948, the introduction of the new Deutsche mark would stabilize the economy and soon lead to the *Wirtschaftswunder* of the 1950s. Almost as miraculously, the number of animals killed in the moors went down abruptly, indicating at least that the slashes had not only come from a wolf or a dog gone wild. In the twilight of an evening at the end of August, though, a farmer would finally shoot the purported "strangler" dead. Proudly he presented the lifeless body that measured almost six feet long and weighed ninety-nine pounds to the cameras of the new newsreel *Welt im Film* (World in film). The autopsy did not produce any clear result as to whether it was a wild wolf, a dog-wolf hybrid, or a wolf that had been raised in captivity.[31]

Be that as it may, the "strangler" was dead, danger averted, "and the invented mythical creature disappeared from the overheated imagination of the people. In the bogs of Lichtenmoor, the livestock grazes again in ruminative peace."[32] So the ending went for the piece from *Welt im Film*, the Allies' answer to the Nazis' *Deutsche Wochenschau*. It was created by the British and the Americans with the goal of "reeducation," to "liberate" the Germans once and for all from their brown-shirted beliefs. And so it would almost seem that, along with the "strangler," the beasts of the recent past had been eradicated at the same time too—at least on the movie screen at the cinema.

ACKNOWLEDGMENTS

More than two years of work went into this book. The idea for it came to me in about 2015, and my agent Thomas Hölzl, who has always stood by my side with sound advice, supported it. I thank the Hanser Verlag and my editor, Annika Domainko, for taking part in this expedition to explore the fauna of the Third Reich. Moreover, I want to thank Nicola von Bodman-Hensler, who helped get this book off the ground.

I would also like to thank the University of Wisconsin Press, including Amber Cederström, Sheila McMahon, and their team, for their confidence in bringing my book to Anglophone audiences. The fact that it reads so well in English is thanks to the terrific work of John R. J. Eyck, who creatively rendered this translation with much love for the detail and sense of style in my work.

I owe special thanks to Heinrich Schlange-Schöningen, who made unpublished material on the history of his family and on life at the former estate of Schöningen in Pomerania available to me. In addition, numerous people helped me with their knowledge and their valuable guidance. I am grateful for the insights of Daniel Baranowski, Wilhelm Bode, Hartmut Böttcher, Andreas Gautschi, Rikola Gunnar-Lüttgenau, Christoph Hinkelmann, Ernst Kalm, Wolfgang Matz, Theresia Mohnhaupt, Thomas Paulke, Werner Philipp, Joachim Radkau, Mieke Roscher, Wolf Stegemann, Sabine Stein, Helmut Suter, Frank Uekötter, Julia Voss, and Albrecht Weber. For their helpful corrections to the manuscript, I also want to thank Michael Kazmierski, Bruno Treu, and Jutta Wallerich.

Last but not least, I especially thank my wife, Juliane, my first and foremost editor and critic, for her perspective when I—surveying all the facts and figures on Hitler's hounds and Goering's harts and hinds—would at times lose mine.

NOTES

Prologue

1. "Zerstreuung und Unterhaltung" and "Tiere in ihrer Schönheit und Eigenart vorzuführen, die sie sonst in freier Wildbahn zu beobachten und kennen zu lernen kaum Gelegenheit haben" (Archiv der Gedenkstätte Buchenwald, NS 4 Bu 33, Film 3).

2. "Wir wollten den Zoo wieder sichtbar machen" and "Es ist irritierend, sich vorzustellen, wie die Nazis mit ihren Kindern den Zoo besuchten und Tiere beobachteten, während nebenan Menschen starben. Weil man erkennt, dass ein Teil der eigenen Normalität, wie eben ein Zoo, auch zu einer Welt gehören kann, der man sich überhaupt nicht zugehörig fühlt" (personal communication with Rikola-Gunnar Lüttgenau, January 2019).

3. "Die SS hat es sich schön gemacht" (personal communication with Rikola-Gunnar Lüttgenau, February 2019).

4. Kogon 1947, 303; Hackett 2002, 164. A copy of the picture book by Kurt Dittmar, *Bärenjagd in Buchenwald*, is held the Archiv der Gedenkstätte Buchenwald (9962). For newspaper articles, see Stange 2015 and Holtz 2018.

5. Official statement of concentration camp survivor Leopold Reitter (Archiv der Gedenkstätte Buchenwald, 31/98).

6. "Ich habe immer wieder Überlebende, die zu Gedenkveranstaltungen hier zu Besuch waren, danach gefragt. Aber an ein Nashorn konnte sich keiner erinnern." (conversation with Sabine Stein, Buchenwald, February 2019).

7. For images from the camp zoo, see Yad Vashem Photo Collections, https://photos.yadvashem.org.

8. On October 5, 1938, the archives of the Leipzig zoo recorded the release of a female brown bear to the Weimar Buchenwald concentration camp (personal communication from Jana Ludewig, Leipzig Zoo Archives, April 2, 2019).

9. Archiv der Gedenkstätte Buchenwald, 31/106597.

10. Archiv der Gedenkstätte Buchenwald, NS 4 Bu 102, Film 8.

11. "Deshalb ging die SS offenbar davon aus, dass sie 'von Natur aus' besonders gut mit diesen Tieren umgehen konnten" (personal communication with Lüttgenau, February 2019).

12. Personal communication with Lüttgenau, February 2019.

13. Archiv der Gedenkstätte Buchenwald, 018.094.

14. Lüttgenau 1993, 15–16.

15. "jegliches Füttern und Necken" (Archiv der Gedenkstätte Buchenwald, NS 4 Bu 33, Film 3).

16. "Wenn ein Junges eingeht, hart bestrafen" (Archiv der Gedenkstätte Buchenwald, NS 4 Bu 102, Film 8).

17. Longerich 2013, 309 ("Ob bei dem Bau eines Panzergrabens 10 000 russische Weiber an Entkräftung umfallen oder nicht, interessiert mich nur insoweit, als der Panzergraben für Deutschland fertig wird. Wir werden niemals roh und herzlos sein, wo es nicht sein muss: das ist klar. Wir Deutsche, die wir als einzige auf der Welt eine anständige Einstellung zum Tier haben, werden ja auch zu diesen Menschentieren eine anständige Einstellung einnehmen" [Longerich 2008, 320]).

18. Höß 2006, 32.

19. Hoess 1951, 172 ("Ich mußte den Vernichtungsvorgang, das Massenmorden weiter durchführen, weiter erleben, weiter kalt auch das innerlich zutiefst Aufwühlende mit ansehen"and "Hatte mich irgendein Vorgang sehr erregt, so war es mir nicht möglich, nach Hause zu meiner Familie zu gehen. Ich setzte mich dann aufs Pferd und tobte so die schaurigen Bilder weg oder ich ging oft des Nachts durch die Pferdeställe und fand dort bei meinen Lieblingen Beruhigung" [Höß 2006, 199–201]).

20. For the Reichstierschutzgesetz, see Reichsministerium de Innern 1933, 1:987.

21. Klueting 2003, 83–85. Hitler himself, however, said, "We ought not go so far as wanting to position animals as better than humans" ("Man dürfe nicht so weit gehen, die Tiere etwa besser stellen zu wollen als den Menschen" [Klueting 2003, 85]).

22. "integraler Bestandteil der Neuordnung der Gesellschaft auf völkisch-rassistischer Grundlage" (Möhring, 2011, 230).

23. According to Eugen Kogon (1947, 303), it gave him "Neronian satisfaction" ("neronisches Vergnügen").

24. "Noch im Jahre 1944, als im Lager große Hungersnot herrschte, bekamen die Raubvögel, Bären und Affen täglich Fleisch, das selbstverständlich aus der Häftlingsküche genommen und so der Verpflegung der Häftlinge entzogen wurde" (Archiv der Gedenkstätte Buchenwald, 31/98).

25. "weil man befürchtet, dass der Fokus auf die Tiere zu einer Bagatellisierung der menschlichen Opfer führe" (interview with Mieke Roscher, June 28, 2019).

26. Adorno 1974, 105 ("Entrüstung über begangene Grausamkeiten wird um so geringer, je unähnlicher die Betroffenen den normalen Lesern sind" and "Vielleicht ist der gesellschaftliche Schematismus der Wahrnehmung bei den Antisemiten so geartet, daß sie die Juden überhaupt nicht als Menschen sehen. Die stets wieder begegnende Aussage, Wilde, Schwarze, Japaner glichen Tieren, etwa Affen, enthält bereits den Schlüssel zum Pogrom. Über dessen Möglichkeit wird entschieden in dem Augenblick, in dem das Auge eines tödlich verwundeten Tiers den Menschen trifft. Der Trotz, mit dem er diesen Blick von sich schiebt—'es ist ja bloß ein

Tier'—wiederholt sich in den Grausamkeiten an Men schen, in denen die Täter das 'nur ein Tier' immer wieder sich bestätigen müssen, weil sie es schon am Tier nie ganz glauben konnten" [Adorno 1951, 133]).

Chapter 1. Blood Ties

1. Sandner 2016, 126. A number of sources use the term "Fuchsl" instead of "Foxl" (translator's note: "Foxl" is a hybrid Anglo-Bavarian variation on "Foxie").

2. Sandner 2016, 164–70.

3. Kershaw 1998, 93.

4. Mohnhaupt 2017, 16.

5. "Uns hat in den Tagen, da die Welt wankte, ein liebes Hundebellen mehr gegeben, als die klügsten Worte der Menschen" (Hohlbaum 1932, 190).

6. Weinberg 2008, 177–78 (cf. Jochmann 1980, 219–20); Kershaw 1998, 75–76.

7. Weinberg 2008, 177–78 (cf. Jochmann 1980, 219–20).

8. Longerich 2016, 125.

9. "Besprechung mit anschließendem Frühstück" and "mit der Endlösung der Judenfrage zusammenhängende Fragen" (Haus der Wannsee-Konferenz 2021). See also Heydrich's introduction to Martin Luther (January 8, 1942) in addition to the official record of the January 20, 1942, Wannsee Conference.

10. On the Wannsee Conference and the competition between Himmler and Heydrich as well as on Hitler's role in the conference, see Longerich 2016, 161–63, and Jäckel 1992, n.p.

11. Weinberg 2008, 198 ("Am besten, sie gehen nach Rußland. Ich habe kein Mitleid mit den Juden" [Jochmann 1980, 241]). A more literal translation is "At best, they'll go to Russia. I have no sympathy with the Jews."

12. For a more specific survey of Hitler's hounds, see Wohlfromm and Wohl-fromm 2001, 178–86.

13. Weinberg 2008, 188 (cf. Jochmann 1980, 231).

14. Fest 1974, 749 ("lieber ein toter Achill als ein lebender Hund" [Fest 2013, 1057]).

15. Möhring 2011, 243.

16. Wippermann 1998, 195.

17. Roscher 2016, 30.

18. On early domestication, see Fuhr 2016, 158, and Barth 2014, 201; for "self-domestication," see Reichholf 2014, 28–30, and Fuhr 2016, 160.

19. Barth 2014, 201.

20. Verein für Deutsche Schäferhunde 1999, 19–21.

21. Wippermann and Berentzen 1999, 70.

22. Wippermann and Berentzen 1999, 69.

23. von Stephanitz 1923, 110 ("Angehörigen einer allgemeinen, weit verbreit-eten Rasse" [von Stephanitz 1921, 107]); Räber 2001, 260.

24. Zelinger 2018, 322.

25. von Stephanitz 1923, 136–37 ("Mit kräftigen Knochen, schönen Linien und edel geformtem Kopf," "das Gebäude trocken und sehnig," "der ganze Hund

ein Nerv," "Wundervoll in seiner anschmiegenden Treue zum Herrn," and "allen anderen gegenüber eine rücksichtslose Herrennatur" [von Stephanitz 1921, 132]).

26. von Stephanitz 1923, 136 (cf. von Stephanitz 1921, 131).

27. von Stephanitz 1923, 134–35 (cf. von Stephanitz 1921, 129–31).

28. Verein für Deutsche Schäferhunde 1999, 21.

29. von Stephanitz 1923, 49 ("nur bissige und scheue Hunde" [von Stephanitz 1921, 48]).

30. von Stephanitz 1923, 50 ("auszeichnenden Eigenschaften beider Elternrassen" [von Stephanitz 1921, 48]). See also Darwin 1868, 192.

31. Barth 2014, 205–6.

32. Denk 2011, 89.

33. Haeckel 1904, 135.

34. Barth 2014, 213; Wunder n.d.

35. Wippermann and Berentzen 1999, 64–65; Verein für Deutsche Schäferhunde 1999, 20–21.

36. "humane Gefühlsduseleien" and "hindern oder verzögern nur die Wirksamkeit der natürlichen Zuchtwahl" (Ploetz 1895, 147).

37. Barth 2014, 207.

38. Reichsministerium des Innern 1933, 1:529.

39. von Stephanitz 1923, 383 ("Wir können unsere Schäferhundzucht recht wohl mit der menschlichen Gesellschaft vergleichen" [von Stephanitz 1921, 397)]. On whelps, see von Stephanitz 1923, 428–29 (cf. von Stephanitz 1921, 432–33).

40. Wippermann 1998, 195.

41. Reichsministerium de Innern 1935, 1:1146.

42. Wippermann and Berentzen 1999, 68.

43. Barth 2014, 202.

44. Wippermann and Berentzen 1999, 60–61.

45. Verein für Deutsche Schäferhunde 1999, 25.

46. On Great Danes, see Wippermann and Berentzen 1999, 48, and Verein für Deutsche Schäferhunde 1999, 19; for dachshunds, see Wippermann and Berentzen 1999, 59.

47. Zieger 1973, 473–75; Roscher 2018, 76–77.

48. Weinberg 2008, 189 (cf. Jochmann 1980, 231); Wippermann and Berentzen 1999, 82.

49. "Der Werwolf hält selbst Gericht und entscheidet über Leben und Tod" (Ahne 2016, 42).

50. "Niemanden haßt der Hund so wie den Wolf" and "er erinnert ihn an seinen Verrat, sich dem Menschen verkauft zu haben—daher er dem Wolf seine Freiheit neidet" (Tucholsky 2013, 29).

51. "Wir kommen als Feinde! Wie der Wolf in die Schafherde einbricht, so kommen wir" (Goebbels 1935, 73).

52. "alle körperlichen und moralischen Verfallserscheinungen mit den Domestikationserscheinungen der Haustiere wesensgleich sind" (Lorenz 1942, 300–301).

53. Brehm 1876, 528–29.

54. Strombeck 1829, 334–35.

55. Brehm 8176, 528.

56. Fuhr 2016, 19–22 (translator's note: from 1871 to 1918, Alsace was annexed to the Wilhelminian Empire and so was for all intents and purposes Prussian territory).

57. Świętorczecki 1938, 256.

58. Bieger 1940, 173. For the term "nichtjagdbares Haarwild" ("furred game that cannot be hunted"), see Giese and Kahler 1944, 227; on the purported protected status of such game, see Möhring 2011, 239, and Sax 2017, 64, 66. The wolf was subject to *freier Tierfang* (at-large animal trapping), which was regulated by the German civil code, and thus had no special protection status.

59. Frevert 1977, 168–73.

60. Sax 2017, 64–67; personal correspondence with Wilhelm Bode, January 15, 2019, and April 24, 2019.

61. "vorzügliche Sanitätspolizei," "Heger des Rotwilds," and "wie er zur Nahrung braucht" (Świętorczecki 1938, 256–57).

62. "Sollte es möglich sein, den Wolf an Fütterungen zu gewöhnen, besonders im Winter, so sollte es vielleicht auch für ein Land mit hoher Kultur möglich sein, sich wenigstens einen kleinen Bestand an Wölfen zu leisten" (Bieger 1940, 173).

63. Dirscherl 2012, 146; Wippermann and Berentzen 1999, 82.

64. Eberle and Uhl 2005, 158; Junge 2011, 74; Sontheimer 2015, n.p.

65. Kellerhoff 2017, n.p.; Misch 2010, 105–6.

66. Misch 2010, 106.

67. Peuschel 1982, 20; Weinberg 2008, 189 (cf. Jochmann 1980, 231).

68. Misch 2010, 106.

69. Wohlfromm and Wohlfromm 2001, 183.

70. Junge 2011, 64.

71. "Blondi, mach Schulmädchen!" "Blondi, sing," and "Blondi sing tiefer, wie Zarah Leander!" (Junge 2011, 105).

72. Reichsministerium für Volksaufklärung und Propaganda 1942, 6:30–39.

73. Wippermann and Berentzen 1999, 83.

74. "ihre Rache" (Zoller 1949, 100).

75. Speer 2008, 300 ("wichtiger als selbst seine engsten Mitarbeiter" [Speer 1969, 312]).

76. Goebbels 1948, 138 ("Dieser Hund darf sich in seinem Bunker alles erlauben," and "Er ist im Augenblick derjenige, der ihm am nächsten steht" [Goebbels 1977, 39]).

77. Speer 2008, 301 (cf. Speer 1969, 313).

78. Schroeder and Joachimsthaler 1985, 130–31.

79. "Was haben Sie mit meinem Hund gemacht," "Sie haben mir das einzige Wesen, das mir wirklich treu ist, abspenstig gemacht," "Ich lasse den Hund erschießen!" (Sauerbruch 1979, 427–29).

80. Perz 1996, 139–43; Höß 2006, 276–77.

81. Perz 1996, 140–42.

82. Räber 2001, 28–31.

83. Glazar 1983, 71.

84. Benz and Distel 2005, 1:37.

85. Benz 2011, 299–301.

86. Glazar 2012, 146–47.

87. "Mensch, fass' diesen Hund!" (Rückerl 1977, 230, 235); Klee, Dreßen, and Rieß 1988, 225.

88. "Hunde, die an der Außenseite der Lager revieren, müssen zu derartig reißenden Bestien erzogen werden, so wie es die Hetzhunde in Afrika sind"and "Sie müssen so abgerichtet sein, daß sie mit Ausnahme ihres Wärters jeden anderen zerreißen. Dementsprechend müssen die Hunde gehalten werden, damit kein Unglück passieren kann. Sie sind eben nur bei Dunkelheit heranzulassen, wenn das Lager geschlossen ist und müssen morgens wieder eingefangen werden" (Perz 1996, 144–45).

89. Höß 2006, 183.

90. Wippermann and Berentzen 1999, 78.

91. Josselin 2017, 20:32–36.

92. Perz 1996, 152; Roscher 2016, 36.

93. "ab sofort sämtliche unter den Diensthunden befindlichen Boxerhündinnen nur noch von Airedaleterrier- und Schäferhund-Rüden gedeckt [werden dürfen]" (Perz 1996, 152).

94. "an ausreichender Intelligenz, an wirklichem Interesse sowie an echter Liebe zum Tier" (Perz 1996, 147–48).

95. Höß 2006, 181–82; Perz 1996, 152.

96. Pröse 2016, 208.

97. Schmoller 2005, 30–33; Freund 1989, 354–56.

98. Lower 2013, 257.

99. Wippermann and Berentzen 1999, 79–80; Höß 2006, 181.

100. Russell 2008, 206; Lower 2013, 244.

101. Benz and Distel 2005, 1:254; see also Glazar 1983, 72.

102. Klee, Dreßen, and Rieß 1988, 224–25; Münster 2000, 183.

103. On Barry, see Rückerl 1977, 235–37, and Klee, Dreßen, and Rieß 1988, 224–25 and 257–58.

104. Taschwer 2015, n.p.

105. "viel feinfühliger als reinrassige Tiere" and "das Spiegelbild des Unterbewusstseins seines Herrn" (von Sagel 1981, 57); "in der Verhaltensphysiologie anerkannt, dass derselbe Hund zeitweilig brav und harmlos, zeitweilig auch gefährlich und bissig sein könne. . . . Er passe sich eben ganz den Stimmungen und Launen seines Herrn an. Wenn ein Hund eine neue Hund-Herren-Bindung eingehe, könne sich sein Charakter so gar völlig wandeln" (Rückerl 1977, 237).

106. "wegen gemeinschaftlichen Mordes an mindestens 300000 Personen und wegen Mordes in 35 Fällen an mindestens 139 Personen" and "satanische Grausamkeit" (von Sagel 1981, 215).

107. Thadeusz 2011, n.p.

108. Münster 2000, 180; Thissen 2014, n.p.

Chapter 2. Digestive Affinities

1. Falkenberg and Hammer 2008, 322; Zorn1963, 38–39.

2. Wuketits 2011, 91.

3. Gerriets 1933, 329. On the domestic pig as a farm animal, see Zorn 1963, 18; on the number of self-sufficient people, see Rahlf 2015, 42; on slaughtering pigs, see Saraiva 2018, 116.

4. Thünen-Institut 2017, 2018.

5. Kalm 1996, 81.

6. Hecht 1979.

7. H. Schlange-Schöningen 1946, 17; Schlange-Schoeningen 1948, 24.

8. "Schöninger Schlange den Kopf zu zertreten," playing on the literal meaning of the last name "Schlange," that is, "snake" (Gies 2019, 597).

9. "Quelle der deutschen Wiedergeburt" (Gies 2019, 396).

10. H. Schlange-Schöningen 1946, 5; Schlange-Schoeningen 1948, 13.

11. For Schlange-Schöningen's biography, see Trittel 1987, 27–32, and Gies 2019, 395–97.

12. Trittel 1987, 25 (translator's note: 750 hectares equals roughly 1,850 acres).

13. E. Schlange-Schöningen n.d.

14. Ramm 1922, 17; H. Schlange-Schöningen 1947, 145.

15. E. Schlange-Schöningen n.d.

16. Hecht 1979.

17. Hecht 1979.

18. H. Schlange-Schöningen 1946, 140; Schlange-Schoeningen 1948, 178.

19. Gerhard 2015, 24; Dornheim 2011, 18; Pollmer 2015, n.p.

20. "britisch-jüdische Verschwörung" (Dornheim 2011, 18); Sax 2017, 59.

21. Radkau 2002, 296; Sax 2017, 66.

22. Gies 2019, 35–37; Hutter 2016, n.p. (translator's note: a *Vordiplom* is approximately the same as a two-year associate's degree in the U.S. higher education system).

23. Corni and Gies 1997, 24.

24. Ammon 1906, 40–41; Spengler 1928, 96 ("den ewigen Menschen" and "die immer fließende Quelle des Blutes" [Spengler 1998, 669]).

25. Gerhard 2012, 116–17.

26. Corni 1993, 20.

27. Longerich 2008, 137–38; for a critical commentary, see Corni 1993, 20–21.

28. Corni 1993, 21.

29. "eifrigsten Ideologen der Bewegung" and "Blut und Boden" (Longerich 2008, 137).

30. "rassische Avantgarde" (Longerich 2008, 762).

31. Longerich 2008, 138.

32. Peuschel 1982, 133; Kater 2014, 577.

33. "kultischen Unfug" (Goebbels 1992, 877).

34. Peuschel 1982, 136; see also Longerich 2008, 118.

35. Peuschel 1982, 139; Ackermann 1993, 107.

36. Ackerman 1993, 107 ("rein äußerlich die Menschen abzusieben" [Ackermann 1989, 126–27]).

37. Saraiva 2018, 103–4.

38. Corni and Gies 1997, 25.

39. Reischle 1935, 29.

40. "Kein Opfertier ist in seiner Deutung so umstritten, kein Haustier schwankt so zwischen völligem Abgelehnt und höchstem Verehrtwerden" (Darré 1933, 5).

41. Darré 1933, 14.

42. "Während das Hausschwein uns über die nordischen Völker die klare Auskunft gibt, daß sie Siedler gewesen sein müssen, beweisen die Semiten mit ihrer Ablehnung alles dessen, was mit dem Schwein zusammen hängt, ebenso klar ihr Nomadentum" (Darré 1933, 15).

43. "faunistischer Antipode jeden Wüstenklimas" (Darré 1933, 17).

44. "parasitär" (Darré 1933, 10).

45. Falkenberg and Hammer 2006, 57.

46. Harris 1998, 74.

47. Harris, 1998, 68.

48. Bornemann 1953, 7.

49. "zu den unbekannten Helden der Umweltgeschichte" and "Schweinekriege" (Radkau 2002, 76–77). On the *Schweinekrieg*, see, e.g., Seebo 1998, 5–13.

50. Zorn 1963, 27. On the integration of pigs into agriculture, see Ramm 1922, iii.

51. Saraiva 2018, 124.

52. Radkau 2002, 295.

53. "Der totale Krieg ist nicht nur eine Frage der Waffenentscheidungen, sondern mit in erster Linie eine solche der gesicherten Volksernährung" (Darré 1937, 9–10).

54. Wowra 1933, 1.

55. Sax 2017, 59.

56. Saraiva 2018, 130.

57. Saraiva 2018, 123.

58. Berg 2015, 472; Schmitz-Berning 2000, 210.

59. Spiekermann 2018, 390.

60. Köstering and Rüb 2003, 140.

61. "die bei ihnen anfallenden Küchen- und Nahrungsmittelabfälle dem Ernährungshilfswerk zur Verfügung zu stellen und die Abfälle in die dafür aufgestellten Haussammeleimer zu schütten" (Reichsministerium de Innern 1939, 1:2104).

62. A photograph of one of the NSV pig signs from 1936 is available in the NSKOV and NSV Collection, SZ Photo/Süddeutsche Zeitung Photo, image ID 00335353, https://www.sz-photo.de/id/00335353.

63. Berg 2015, 463–65.

64. H. Schlange-Schöningen 1946, 138–39; Schlange-Schoeningen 1948, 178.

65. "ernährungspolitischen Gesichtspunkten" (Longerich 2008, 435).

66. Longerich 2008, 428–29.

67. Weinberg 2008, 33 ("Ostwall . . . aus lebenden Menschen" [Jochmann 1980, 68]).

68. Gies 2019, 666.

69. "Auch ich erlitt den Heldentod / ich starb an Darrés Gerstenschrot!" (Radkau 2002, 296).

70. Schlange-Schoeningen 1948, 178 ("Millionen Zentner Kartoffeln und Rüben sind erfroren" and "Wegen Futtermangels ein Schweinemord, der den . . . des kleinen Weltkriegs bei weitem übertrifft. Wird es überhaupt möglich sein, den Anschluß an die nächste Ernte zu erreichen. Von Weltanschauungspropaganda wird kein hungernder Magen satt" [H. Schlange-Schöningen 1946, 155]).

71. Corni and Gies 1997, 479.

72. Münkel 1996, 377.

73. "Der deutsche Schweinehalter kennt seine erste Pflicht: Durchhalten und siegen helfen durch Sicherung der Volksernährung!" (Editors 1941, 1); "Wer Brotgetreide verfüttert, hilft dem Feind!" (Schneider 1941, 2).

74. "die verstärkte Herausnahme von Schweinen" (Corni and Gies 1997, 480).

75. "ein Eingriff in die Substanz des Hofes zu Lasten der Zukunft" (Backe 1943, 1).

76. On the self-sufficient population, see Gerhard 2015, 54; on slaughter permissions, see Corni and Gies 1997, 494.

77. "kriegsschädlichem Verhalten" and "Wer Rohstoffe oder Erzeugnisse, die zum lebenswichtigen Bedarf der Bevölkerung gehören, vernichtet, beiseiteschafft oder zurückhält und dadurch böswillig die Deckung dieses Bedarfs gefährdet, wird mit Zuchthaus oder Gefängnis bestraft. In besonders schweren Fällen kann auf Todesstrafe erkannt werden" (Reichsministerium de Innern 1939, 1:1609).

78. Münkel 1996, 384–87.

79. "kollektiv verübten Delikt" (Herlemann 1993, 309).

80. Rath 2018, n.p.

81. Junge 2011, 78; Zoller 1949, 78.

82. Zoller 1949, 78.

83. Fest 1974, 535 (cf. Fest 2013, 764).

84. Speer 2008, 301 ("Leichentee" [Speer 1969, 313]) (translator's note: in Hitler's reference here to animals, however, the more fitting translation is "carcass tea").

85. "Widerwillen gegen Fleisch" and "der Fleischgenuß den Menschen vom natürlichen Leben entferne" (Zoller 1949, 77).

86. Weinberg 2008, 89 (cf. Jochmann 1980, 128).

87. Weinberg 2008, 430 (cf. Picker 1976, 432).

88. Weinberg 2008, 176 (cf. Jochmann 1980, 217–18).

89. E. Göring 1967, 143–44.

90. Weinberg 2008, 89 (cf. Jochmann 1980, 57, 127).

91. "Fleischgenuss für die künftigen Generationen einschränken" and "Fleisch und die Wurstwaren durch ebenso wohlschmeckendes, den Gaumen und das Bedürfnis des Körpers Befriedigendes" (Longerich 2008, 347).

92. "warme Abendkost in Gestalt von Suppe, Pellkartoffeln und dann kalter Zugabe" (Longerich 2008, 347).

93. "strengstens zu verbieten" (Longerich 2008, 48).

94. Brucker 2015, 216.

95. "oberstes Gebot der Sittlichkeit" (Schwantje 1928, 8).

96. "Wer das Verhalten der Tiere gründlich beobachtet und vorurteilsfrei deutet, es als ein schweres Unrecht erkennen, Tiere nur deshalb schlachten zu lassen, um sich einen Gaumengenuss zu verschaffen" (Schwantje 1942, 11).

97. "Es würde bestimmt keine wirtschaftliche Krise eintreten, wenn heute Tausende von Europäern sogleich zur vegetarischen Ernährung übergingen. Im Gegenteil, jede Einschränkung der Fleischproduktion wird auch auf die Volkswirtschaft einen günstigen Einfluss ausüben" (Schwantje 1942, 14).

98. Brucker 2015, 215.

99. Brucker 2015, 219.

100. "Warmblütige Tiere sind beim Schlachten vor Beginn der Blutentziehung zu betäuben" (Reichsministerium de Innern 1933, 1:203).

101. "offene, brutale und unbeirrbare Grausamkeit des Juden" ("Ein erschütterndes Filmdokument" 1938).

102. "humanes Schlachten" (Jentzsch 1998, 48).

103. Klueting 2003, 79; Dirscherl 2012, 37–38.

104. Klueting 2003, 81; Jentzsch 1998, 45.

105. Jentzsch 1998, 48.

106. Jentzsch 1998, 44.

107. Motadel, Held, and Hornung 2017, 306–7.

108. Giese and Kahler 1944, 195.

109. Giese and Kahler 1944, 210–11.

110. "Jeder am Tierversand Beteiligte hat die Pflicht, mit dafür zu sorgen, daß die Tiere die Bestimmungsstation in bester Verfassung erreichen" (Giese and Kahler 1944, 197).

111. "Verhindert Transportverluste," 1940, 238.

Chapter 3. Drawing the Curtain on Larval Stages

1. Reichsverband Deutscher Kleintierzüchter 1937, 38–48.

2. Oberthür 2002, 28–29.

3. Reichsministerium für Wissenschaft, Erziehung und Volksbildung 1943, 225.

4. Reichsministerium für Wissenschaft, Erziehung und Volksbildung 1940, 7.

5. Schmitz-Berning 2000, 210–11.

6. "die Ziele des Vierjahresplans erfüllen zu helfen" (Reichsministerium für Wissenschaft, Erziehung und Volksbildung 1938, 254).

7. "die Wirtschaftsblockade unserer Gegner im gegenwärtigen Kriege erfordert einen verstärkten Einsatz der Schulen" (Reichsministerium für Wissenschaft, Erziehung und Volksbildung 1940, 7).

8. Reichsministerium für Wissenschaft, Erziehung und Volksbildung 1940, 7–8.

9. Reichsverband Deutscher Kleintierzüchter 1937, 16–19.

10. Reichsministerium für Wissenschaft, Erziehung und Volksbildung 1935, 464; Wylegalla 2011, n.p.

11. Reichsministerium für Wissenschaft, Erziehung und Volksbildung 1943, 225; Reichsverband Deutscher Kleintierzüchter 1937, 76.

12. "Unerbittlichkeit der Natur" (Reichsministerium für Wissenschaft, Erziehung und Volksbildung 1940, 8).

13. "Reinheit der Rasse" (Bäumer-Schleinkofer 1992, 49).

14. On silkworm diseases, see Reichsverband Deutscher Kleintierzüchter 1937, 49–51.

15. Weinberg 2008, 269 ("nur ein allgemeines Wissen geben" and "Was braucht ein Junge, der Musik üben will, Geometrie, Chemie, Physik? Was weiß er davon später noch! Nichts! Das ganze Detaillierte soll man lassen!" [Picker 1976, 119]) (translator's note: it is clear that the Weinberg edition embellishes here, given that the second half of this diatribe reads: "Nothing! They should leave all that detailed knowledge be!").

16. "Das Schwache muss weggehämmert werden," "Ich werde sie in allen Leibesübungen ausbilden lassen. Ich will eine athletische Jugend. Das ist das Erste und Wichtigste. So merze ich die Tausende von Jahren der menschlichen Domestikation aus. So habe ich das reine, edle Material der Natur vor mir. So kann ich das Neue schaffen," and "eine Jugend, vor der sich die Welt erschrecken wird" (Grün 1979, 100); for "Das Schwache muss weggehämmert werden," see also Picker 1976, 97. The English translation of *Hitler's Table Talk* edited by Weinberg does not include Hitler's remarks on how to mold young people.

17. Flessau 1977, 20.

18. Struck 2015, n.p.

19. Flessau 1977, 19; "Welche Wahrscheinlichkeit besteht für das Auftreten eines jüdischen Merkmals aus einer Ehe zweier Mischlinge 2. Grades?" (Blumesberger 2009, 2).

20. "Nationalsozialistisches Denken muß biologisches Denken sein" (Bäumer-Schleinkofer 1992, 51).

21. "wissenschaftlich belegt" and "alle Kultur rassisch bedingt ist" (Reichsministerium für Wissenschaft, Erziehung und Volksbildung 1938, 91).

22. "Überlegenheit der nordischen Rasse" and "völkische Bedeutung" (Flessau 1977, 80–81).

23. Bäumer-Schleinkofer 1992, 56.

24. Flessau 1977, 11–13; "Du bist nichts, Dein Volk ist alles!" (Bäumer-Schleinkofer 1992, 52).

25. "Wie selten ist es uns in der Schule beschieden, Fleiß, Treue und Gewissenhaftigkeit so schnell in greifbare Ernte umgesetzt zu sehen wie hier" and "Bei aller Arbeit hat sich so manches Kind, das sonst nie in Erscheinung tritt, als brauchbares Glied der Gemeinschaft erwiesen" (Reichsministerium für Wissenschaft, Erziehung und Volksbildung 1940, 8).

26. Reichsverband Deutscher Kleintierzüchter 1937, 16–19.

27. Leimann 2016, 9.

28. Weiser 2003, n.p.

29. "verpflichtet, in der Hitler Jugend Dienst zu tun" (Reichsministerium de Innern 1939, 1:710).

30. Laue 2015, 40.

31. "In unseren Augen, da muss der deutsche Junge der Zukunft schlank und rank sein, flink wie Windhunde [*sic*], zäh wie Leder und hart wie Kruppstahl" (Stiftung Haus der Geschichte 2011, n.p.).

32. Laue 2015, 40.

33. "einzigartigen Waffentaten" and "härtesten Kämpfen gegen stärkste feindliche Übermacht" (Reichsministerium für Volksaufklärung und Propaganda 1941b, 01:22).

34. "Wir wissen nur eines, wenn Deutschland in Not, / Zu kämpfen, zu siegen, zu sterben den Tod" (Reichsministerium für Volksaufklärung und Propaganda 1941b, 01:55–02:12).

35. "Jungvolkjungen sind hart, schweigsam und treu, / Jungvolkjungen sind Kameraden / des Jungvolkjungen höchstes ist die Ehre" (Seidel 2008, n.p.). For views on death and on being a soldier, see Rackow 2013 and Freytag 2013.

36. Reichsjugendführung 1938, 17.

37. Hermand 1994, 58–59.

38. Boberach 1982, 35.

39. "selbstloser Einsatz" (Reichsministerium für Volksaufklärung und Propaganda 1941b, 01:39); "schwerer Kampf" and "stolzer Sieg" (02:12–15).

40. Fröhlich 2013, 81–82.

41. "Jeder halte ständig Wacht, / sei bereit zur Abwehrschlacht! / Nur vereint kann es gelingen, / diesen Schädling zu bezwingen" (Köstlin 1941, 6)

42. Herrmann 2009, 89–91; Brümmer 1877, 522–525.

43. Herrmann 2009, 92.

44. "die Furcht vor einer Einschleppung dieses Ungeziefers für grundlos" (Brehm 1877, 186); Brümmer 1877, 522–25.

45. Herrmann 2009, 93; Geißler 1999, 444–46.

46. Industrieverband Agrar 2013, n.p. .

47. Herrmann 2009, 100.

48. Herrmann 2009, 94–95.

49. Geißler 1999, 449.

50. Herrmann 2009, 93.

51. Meibauer 2017, 292–93.

52. Oberthür 2002, 25–27.

53. Herrmann 2009, 95.

54. Geißler 1999, 449; Herrmann, 2009, 96.

55. Klee 2009, 233.

56. Kirchmeier 2013, 74.

57. "die Juden als Gauner und Verbrecher zu erkennen" and "die für die Menschen die gleiche Gefahr bedeuten wie die Wanzen" (Hiemer 1938, 5, 91).

58. "Drohnen gibt es nicht nur bei den Bienen, Drohnen gibt es auch bei Menschen. Es sind die Juden!" (Hiemer 1940, 8).

59. After 1945, the story was censored in part; see Uther 2013, 240 (translator's note: it first appeared in the Grimms' 1815 collection *Kinder- and Hausmärchen*; see Grimm and Grimm 1981, 236–41, and for an English version, see Grimm and Grimm 1884, 97–102).

60. "ewiger Parasit" (Bein 1965, 134).

61. Bein 1965, 125; Stullich 2013, 24.

62. Thierfelder 1979, 12.

63. Bein 1965, 127. Incidentally, the myth of the wandering Jew was not just spread by Nazis. It later found its way into botany as well and is used to this day as the common name for ground-covering plants of the genus *Tradescantia*.

64. Herder 1800, 486 ("als die parasitische Pflanze, die sich beinah allen europäischen Nationen angehängt und mehr oder minder von ihrem Saft an sich gezogen hat" [Herder 1841, 250]). For further explanation of the ostracization of the Jews, see Bein 1965, 127–28.

65. For Hitler's remark ("sich immer weiter ausbreitet"), see Schmitz-Berning, 2000, 462); for Himmler's, see Longerich 2013, 689 (cf. "zersetzende Pest . . . in unserem Volkskörper" [Peterson and Smith 1974, 169]).

66. Himmler 1946, 574 ("Mit dem Antisemitismus ist es genauso wie mit der Entlausung" and "Es ist keine Weltanschauungsfrage, daß man die Läuse entfernt. Das ist eine Reinlichkeitsangelegenheit. . . . Wir sind bald entlaust. Wir haben nur noch 20 000 Läuse, dann ist es vorbei damit in ganz Deutschland" [Peterson and Smith 1974, 201]).

67. Peuschel 1982, 53.

68. "Sie sind hinterlistig, feige und grausam und treten meist in großen Scharen auf. Sie stellen unter den Tieren das Element der heimtückischen, unter irdischen Zerstörung dar. Nicht anders als die Juden unter den Menschen" (Hippler 1940, n.p.).

69. On the historical usage of the term "Schädling," see Jansen 2003, 13–20.

70. "auch der letzten Kuhmagd in Deutschland klargemacht werden, daß das Polentum gleichwertig ist mit Untermenschentum," "immer nur leitmotivartig," "Polnische Wirtschaft," "Polnische Verkommenheit," and "bis jeder in Deutschland jeden Polen, gleichgültig ob Landarbeiter oder Intellektuellen, im Unterbewußtsein schon als Ungeziefer ansieht" (Schmitz-Berning 2000, 622).

71. Klemperer 2006, 15 ("Worte können sein wie winzige Arsendosen: sie werden unbemerkt verschluckt, sie scheinen keine Wirkung zu tun, und nach einiger Zeit ist die Giftwirkung doch da" [Klemperer 1947, 26]).

72. "Wie hinterlasse ich sichtbare Spuren meiner nationalsozialistischen Gesinnung?" and "an Deinen Spuren wird man dich erkennen" ("Rezept," 1934, n.p.).

73. Lieb 2007, 114.

74. "Wie im Herbst 1939 stehen wir nun wieder ganz allein der Front unserer Feinde gegenüber" and "Es ist in den Gauen des Großdeutschen Reiches aus allen waffenfähigen Männern im Alter von 16 bis 60 Jahren der Deutsche Volkssturm zu bilden" (Reichsministerium de Innern 1944, 1:253).

75. Himmler 1943, n.p. ("Fünfzehnjährige machen" and "Denn besser es sterben die fünfzehnjährigen Jungens, als dass die Nation stirbt" (Bayerische Staatsbibliothek" 1943, n.p.).

76. "'Das Gebiet Ostpreußen'—'Hier!' [ruft einer]—'meldet 9482 Kriegsfreiwillige der Hitlerjugend . . . Das Gebiet Köln-Aachen'—'Ja!' [ertönt es, deutlich leiser, von weiter hinten]—'meldet 9715 Kriegsfreiwillige der Hitlerjugend . . . Das Gebiet Moselland . . . meldet 6112 Kriegsfreiwillige der Hitlerjugend'" (Reichsministerium für Volksaufklärung und Propaganda 1944, 12:27–59).

77. "Heute melde ich Ihnen, mein Führer, dass sich von den Hitlerjungen des Jahrgangs 1928 70 Prozent kriegsfreiwillig zu den Fahnen gemeldet haben. Die echte Kriegsfreiwilligkeit dieser Jugend wird in der Kampfmoral auf dem Schlachtfeld lebendig sein" (Reichsministerium für Volksaufklärung und Propaganda 1944, 13:00–13:26).

78. The distinction was awarded for "Tapferkeit vor dem Feind"; see Arnold and Janick 2005, 132.

79. "in Anbetracht der Kriegswichtigkeit des Seidenbaues dafür Sorge zu tragen, daß die Nutzung und Pflege der . . . Maulbeeranpflanzungen in vollem Umfang sichergestellt werden" (Reichsministerium für Wissenschaft, Erziehung und Volksbildung 1945, 16).

80. Geißler 1998, 122; Geißler 1999, 465.

81. Herrmann 2009, 101.

82. Herrmann 2009, 95–96.

Chapter 4. *Morituri*

1. Klemperer 2017, 172.

2. On cats as a purported substitute for children in Eva Klemperer's case, see Zipfel 2014, 41–58.

3. "Wir sind ganz isoliert" and "Unser einziger und getreuester Umgang ist Mucius, genannt Mujel" (Klemperer 2017, 172). In his published diaries, Klemperer wrote of "Muschel" as well as "Mujel."

4. Klemperer 1998, 281 (cf. Klemperer 2015, 1:371).

5. Klemperer 1999c, 51 (cf. Klemperer 2015, 2:58).

6. Klemperer 1998, 277 (cf. Klemperer 2015, 1:539).

7. Klemperer 1999c, 28 (cf. Klemperer 2015, 2:38).

8. Klemperer 1998, 346 (Klemperer 2015, 1:441).

9. Klemperer 2006, 157 (cf. Klemperer, 1947, 179).

10. Klemperer 1998, 234 ("Es ist überhaupt nicht mehr viel Gefühl für die Menschen in mir übriggeblieben. Eva—und dann kommt schon der Kater Mujel" [Klemperer 2015, 1:307]).

11. Klemperer 1999c, 52 ("Der erhobene Katerschwanz ist unsere Flagge" and "wir streichen sie nicht, wir behalten die Nasen hoch, wir bringen das Tier durch" [Klemperer 2015, 2:70]).

12. "tierliebes" (Aly 2005, 351).

13. Möhring 2011, 241; Wippermann and Berentzen 1999, 75.

14. Bundesarchiv 2019, n.p.; Herzig 2010, 59.

15. Walk 2013, 364.

16. "Welch eine niedrige und abgefeimte Grausamkeit gegen die paar Juden" (Klemperer 2015, 2:70) (translator's note: this line has been omitted from the

English edition of the diaries, conceivably conflated with the next sentence, which simply reads: "I feel very bitter for Eva's sake" [Klemperer 1999c, 52]).

17. "Die Haustiere können keine politische Rolle gespielt haben, aber für viele war es das einzige Lebewesen, das sie nach der Zwangsarbeit noch zu Hause freudig begrüßte" (Rewald 1996).

18. Höxter 1997.

19. Schönfeld 1996.

20. "anständige Einstellung zum Tier" (Longerich, 2008 320).

21. Möhring 2011, 234.

22. "unschädlich zu machen" (Reichsministerium de Innern 1936, 1:186).

23. "als 200 Meter vom nächsten bewohnten Hause" (Reichsministerium de Innern 1934, 1:559).

24. "einer der ärgsten nationalistischen Narren" (Hesse and Mann 1999, 148).

25. Vesper 1995, 356. For Vesper's career, see "Vesper, Will" n.d., and Klingenberg 2003. On the topic of hunting, see Vesper 1995, 636.

26. "die Juden unter den Tieren" (Vesper 1995, 356, 636, and 678).

27. Möhring 2007, 179.

28. "unsere hygienischen Helfer bei der Volksgesundung" and "[sie] in keinem Kulturlande so niederträchtig mißhandelt und verfolgt werden wie in Deutschland" (Schwangart 1937, 8).

29. "ein Charakteristikum breiter deutscher Schichten" (Kircher-Kannemann 2017).

30. Petzsch 1964, 60.

31. "sonst hoch entwickelten Tierschutz und einer sonst humanen Tierhaltung" (Schwangart 1937, 15).

32. "abweisender Einsiedler" and "Die Katze besonders ist das Heimtier der Armen, der materiell und seelisch [B]beeinträchtigten, oft genug ihr letztes Glück" (Schwangart 1937, 10–11).

33. Klemperer 1999c, 52 ("Für Eva war es immer ein Halt und ein Trost. Sie wird nun geringere Widerstandskraft haben als bisher" [Klemperer 2015, 2:70]).

34. Klemperer 1999c, 55 ("Du hast immer noch deine Arbeit" and "mir ist alles genommen" [Klemperer 2015, 2:73]).

35. Klemperer dedicates his book *LTI* to Eva (Klemperer 2006, ix; cf. Klemperer 1947, 5).

36. Möhring 2011, 241.

37. Klemperer 1999c, 52 (cf. Klemperer 2015, 2:70).

38. Mackenzie 1960, 65.

39. On domestication, see Schuller 2007, 13; on house cats in Europe, see Fehringer 1953, 94, and Dinzelbacher 2000, 195.

40. "giftigen Atem" (Alcabes 2010, 34).

41. Bluhm 2014, 129–30.

42. Fehringer 1953, 94.

43. Möhring 2011, 239.

44. Zelinger 2018, 255.

45. "niemals ein echtes, deutsches Haustier" and "aus dem Osten eingewanderte Feind" (Zelinger 2018, 254–55).

46. "Je höher ein Volk steht, je bestimmter es sich seßhaft gemacht hat, um so verbreiteter ist die Katze" and "oft rückhaltlos zahm" (Brehm 1876, 463, 351).

47. "auch noch die Montagskater" and "mitversteuert werden" (Zelinger 2018, 167–69) (translator's note: explaining the wordplay in German between a *Kater*, meaning both "tomcat" and "hangover," obviously spoils the joke here.)

48. "geradezu kläglich" (Schwangart 1937, 36).

49. Klemperer 1999c, 54 ("Das Tierchen spielt herum, ist vergnügt und weiß nicht, daß es morgen stirbt" [Klemperer 2015, 2:73]).

50. Klemperer 1999c, 56 (cf. Klemperer 2015, 2:75).

51. Klemperer 1999c, 54 (cf. Klemperer 2015, 2:72). On meat rationing, see Buchheim 2010, 307.

52. Klemperer 1999c, 55 ("Er hat nicht gelitten" [Klemperer 2015, 2:73]).

53. Klemperer 1999c, 56 ("Muschel ist drei Tage zu früh gestorben . . . heute könnte er offiziell der arischen Witwe Elsa Kreidl gehören" [Klemperer 2015, 2:74]).

54. "Die nordische Bewegung ist seit jeher gefühlsmäßig gegen die 'Verhaustierung' des Menschen gerichtet gewesen," "sie kämpft für eine Entwicklungsrichtung, die derjenigen, in der sich die heutige zivilisierte Großstadtmenschheit bewegt, gerade entgegenge setzt ist," and "eine noch schärfere Ausmerzung ethisch Minderwertiger" (Föger and Taschwer 2001, 110, 108).

55. Möhring 2011, 239; Sax 1997, 16; Sax 2017, 118.

56. "ebenso klein wie drollig" (E. Göring 1967, 130).

57. Demandt 2007, 78.

58. Kube 1986, 2; Diels 1949, 63; "Staatsstreich gegen die Wehrmacht" 1974.

59. "Tiere—und Löwen!" (Frevert 1977, 214–16).

60. "fanatischer Tierfreund" (Gritzbach 1938, 104).

61. For the video of Goering with lions at Obersalzberg, see Anonymous 1936. The video is dated 1936, though Goering's automobile accident was recorded in Goebbels's diaries on August 17, 1934 (see Goebbels 2005, 93).

62. Klothmann 2015, 179; Haikal 2003, 158.

63. E. Göring 1967, 130; Knopf and Martens 2004, 54.

64. E. Göring 1967, 131; Rubner 1997, 186.

65. Frevert 1977, 215.

66. Knopf and Martens 2004, 54 and 58; E. Göring 1967, 130.

67. Frevert 1977, 215.

68. Sax 2017, 83.

69. Sax 2017, 21; Spengler 1932, 22 ("Das Raubtier ist die höchste Form des beweglichen Lebens" and "Es gibt dem Typus Menschen einen hohen Rang, daß er ein Raubtier ist" [Spengler 1931, 17]).

70. "Das Schwache muss weggehämmert werden. Stark und schön will ich meine Jugend, herrisch und unerschrocken. Das freie, herrliche Raubtier muss aus ihren Augen blitzen" (Picker 1976, 97). Weinberg's edition of *Hitler's Table Talk* omits this quote, which is interpolated commentary by Picker.

71. Burchardt 2007, 188.

72. Speer 2008, 233 (cf. Speer 1969, 246).

73. On the names for and development of the tiger and panther models, see Burchardt 2007, 185–91, and Höge 2017, 95–107.

74. Klemperer 1999c, 389–96 (cf. Klemperer 2015, 2:536).

75. Klemperer 1999c, 389 (cf. Klemperer 2015, 2:535).

76. Bergander 1977, 320.

77. The figures come from Landeshauptstadt Dresden 2010, n.p.

78. Klemperer 1947, 272.

79. See, among others, Klemperer 1999b, 50, 99, 196–98.

80. Klemperer 2006, 95 ("nicht in Einzelfällen und aus vereinzelter Niedertracht, sondern amtlich und systematisch, und das ist eine der Grausamkeiten, von der kein Nürnberger Prozeß berichtet, und denen ich, läg' es in meiner Hand, einen turmhohen Galgen errichten würde, und wenn mich das die ewige Seligkeit kostete" [Klemperer 1947, 108]).

81. "manchmal weniger von der des dritten unterschieden als etwa das Dresdener Sächsische vom Leipziger" (Klemperer 1999a, 171).

82. Höge 2017, 95–107.

Chapter 5. Raufbold

1. See Ahlborn 1937.

2. "den Schöpfer im Geschöpfe" (Bode 2018, 51–53; Selheim 2009, 165–66; Deutscher Jagdverband 2016, n.p.) (translator's note: the poem's title refers to the greeting exchanged among German hunters, and the first stanza is included on labels for Jägermeister liqueur to this day).

3. On the term "international" here, see Schmitz-Berning 2000, 324–25.

4. "kameradschaftliche Zusammenarbeit der Jäger der ganzen Welt" (H. Göring 1938, n.p.).

5. Fest 1970, 72 ("geboren Nationalsozialist" and "mit einem spontanen Verlangen nach kämpferischer Bewährung sowie einem unreflektierten, elementaren Hunger nach Macht" ([Fest 1963, 104]).

6. Fest 1970, 76–77 (cf. Fest 1963, 111).

7. Kube 1993, 63.

8. Kube 1993, 64–65; Fest 1970, 73–74 (cf. Fest 1963, 106).

9. Gautschi 2010, 65; Peuschel 1982, 70; Kube 1993, 65–66.

10. On the title *Reichsjägermeister*, see Bode 2018, 58, and Gautschi 2010, 64–65.

11. Gautschi 2010, 62–63.

12. Fest 1970, 71 (cf. Fest 1963, 103).

13. Roosevelt and Bullitt 1972, 233.

14. Fest 1970, 78 (cf. Fest 1963, 112).

15. Ahlborn 1937.

16. Bode and Emmert 2000, 152–53.

17. Frevert 1977, 118; Dieberger 2018, 69.

18. Selheim 2009, 162–65.

19. "hemmungslose Promiskuität" (Bode 2018, 38).

20. Perschke 1999, 482.

21. Gautschi 2010, 62–63.
22. Gritzbach 1938, 117.
23. Gautschi 2010, 60.
24. Hasel 1985, 88; Leonhardt 2008, 35.
25. Bode 2016, 58.
26. "öffentliche Ordnung," "gemeine Sicherheit" and "Aasjäger" (Bode and Emmert 2000, 115–20).
27. On the history of organized German hunters at the start of the twentieth century, see Gautschi 2010, 55–57.
28. "politisch Anwalt" (Gautschi 2010, 70–73); Bode and Emmert 2000, 139.
29. On Goering's election, see Bode and Emmert 2000, 141, and Gautschi 2010, 60–62.
30. "Blut lecken" (Gautschi 2010, 60); Dieberger 2018, 68.
31. Gautschi 2010, 60.
32. "Schirmherr der deutschen Jagd" and "sehr schnellen und sehr sicheren Büchsenschützen" ("Quasi-Verrückte" 1994, 257).
33. Bode 2016, 83; Gautschi 2010, 64–65.
34. Bode and Emmert 2000, 146.
35. For the text of the legislation, see Reichsministerium de Innern 1934, 1:549; for its interpretation, see Bode and Emmert 2000, 144–45.
36. Gautschi 2010, 126; Bode 2018, 77, 79.
37. "Die Liebe zur Natur und ihren Geschöpfen und die Freude an der Pürsch in Wald und Feld wurzelt tief im deutschen Volk. Aufgebaut auf uralter germanischer Überlieferung, hat sich so im Laufe der Jahrhunderte die edle Kunst des deutschen Waidwerks entwickelt. . . . Die Pflicht eines rechten Jägers ist es das Wild nicht nur zu jagen, sondern auch zu hegen und zu pflegen. . . . Das Jagdrecht ist unlös bar verbunden mit dem Recht an der Scholle, auf der das Wild lebt, und die das Wild nährt. . . . Treuhänder der deutschen Jagd ist der Reichsjägermeister; er wacht darüber, daß niemand die Büchse führt, der nicht wert ist, Sachwalter anvertrauten Volksguts zu sein" (Reichsministerium de Innern 1934, 1:549).
38. Bode 2016, 98.
39. Personal correspondence with Wilhelm Bode, January 18, 2019.
40. On Frevert's persona, see Bode 2018, 50–55, Dieberger 2018, 69–73, and Hopp and Weitz 2017.
41. Gautschi 2010, 101–2.
42. Frevert 1977, 7.
43. Syskowski 1996, 19; Tautorat 1996, 12; Frevert 1977, 15; Frevert 2007, 135.
44. Bode 2018, 98, 100.
45. Gautschi 2010, 119; Hopp and Weitz 2017; Frevert 1977, 130.
46. Frevert 1977, 165, 174.
47. Frevert 1977, 181–82.
48. Frevert 1977, 168–73.
49. Frevert 1977, 15.
50. Tournier 1972, 193 ("Oger von Rominten" [Tournier 1989, 221]).
51. Gautschi 2010, 118–23.

52. Personal correspondence with Wilhelm Bode, August 13, 2019.

53. Tautorat 1983, 26.

54. The Rominten red deer stock ledger (no. 12, "Raufbold") at the Ostpreußisches Landesmuseum in Lüneburg, Germany, records the death of Raufbold.

55. Neumärker and Knopf 2008, 60, 96–97.

56. Reichsjugendführung 1938, 403; Perschke 1999, 482.

57. Goering's attribution, according to his biographer; see Gritzbach 1938, 116.

58. Weinberg 2008, 73–74 (cf. Jochmann 1980, 111).

59. Speer 2008, 97 ("Wenn wenigstens noch eine Gefahr damit verbunden wäre, wie in den Zeiten, als man mit dem Speer gegen das Wild anging," "Aber heute, wo jeder mit einem dicken Bauch aus der Entfernung sicher das Tier abschießen kann," and "Überreste einer abgestorbenen, feudalen Welt" [Speer 1969, 110]).

60. Weinberg 2008, 73–74 (cf. Jochmann 1980, 111).

61. Gautschi 2010, 163.

62. "reaktionär" (Gautschi 2010, 68).

63. "ihrer verfluchten Jägerei" (Gautschi 2010, 163).

64. "grüne Freimaurer" (Gautschi 2010, 163); Neumärker and Knopf 2008, 97; "best man" and "most loyal Paladin" (Kube 1993, 65, 66).

65. Goebbels 1979, 1 (cf. Goebbels 1977, 55).

66. Goebbels 1979, 2 ("mangelnde Gesinnungstreue" and "wie die Kuh mit der Strahlenforschung" [Goebbels 1977, 56]).

67. "So etwas gab es nicht mal beiden Zaren!" (Kordt 1950, 303).

68. Knopf and Martens 2004, 70–71.

69. Knopf and Martens 2004, 25.

70. On the designation "Nebenaußenminister," see Neumärker and Knopf 2008, 89.

71. Mildner and Resch 1997, 84.

72. Knopf and Martens 2004, 147.

73. "Dem Reichsjägermeister Hermann Goering / Dank für das neue Jagdgesetz / Das deutsche Wild" (Mildner and Resch 1997, 132).

74. "Es ist ein Märchenplatz an dem wir sitzen, denn wir sitzen am Ufer des Werbellin" (Fontane 1910, 494).

75. Gautschi 2010, 104–5.

76. Bode and Emmert 2000, 147.

77. Wohlfromm and Wohlfromm 2017, 178–80; Selheim 2009, 165–66.

78. Frevert 1977, 211; Neumärker and Knopf 2008, 56.

79. "ausgesprochen Zukunftshirsch" (Neumärker and Knopf 2008, 98); Frevert 2007, 221.

80. "Meine Gäste lassen Sie die Kapitalhirsche und mich selbst Ihre Abnormitäten schießen! Herr Reichsjägermeister, dieser Gast war aber auch der Oberbefehlshaber des Heeres" (Frevert 1977, 56–57).

81. "Kapitale und hochkapitale Hirsche werden in Zukunft in den Revieren, in denen ich persönlich während der Hirschbrunft anwesend zu sein pflege, nur durch mich und nicht mehr durch meine Gäste erlegt. . . . Ausnahmen . . . können allein durch mich befohlen werden" (Gautschi 2010, 123).

82. Gautschi 2010, 79.

83. Van Vuure 2005, 64–71.

84. "Urmacher unerwünscht" 1954.

85. Mohnhaupt 2019, 20; on the trophies, see Gautschi 2010, 122.

86. Tautorat 1996, 12; Frevert 1977, 18.

87. Frevert 1977, 17–18.

88. Gautschi 2010, 80.

89. Canetti 1984, 173 ("In keinem modernen Lande der Welt ist das Waldge-fühl so lebendig geblieben wie in Deutschland" [Canetti 1994, 202]).

90. Zechner 2017.

91. On Varus's battle in the Osnabrück region, see Varusschlacht im Osna-brücker Land 2017.

92. Urmersbach 2018, 67–70.

93. Urmersbach 2018, 83–84; Zechner 2017, 4–10.

94. "eingehegte Parke" and "In dieser deutschen Waldfreiheit, die so fremdartig aus unsern übrigen modernen Einrichtungen hervorlugt, liegen mehr bestimmende Einflüsse auf unser höheres Bildungsleben, und namentlich auf die romantische Stimmung in demselben, als mancher sich träumen läßt" (Riehl 1854, 34).

95. "nicht bloß damit uns der Ofen im Winter nicht kalt werde, sondern auch damit die Pulse des Volkslebens warm und fröhlich weiter schlagen, damit Deutsch-land deutsch bleibe" (Riehl 1854, 32).

96. "kommunistischen Gleichmacherei" (Urmersbach 2018, 99).

97. "in hartem Kampfe mit dem Walde schuf sich der deutsche Mensch mit zäher Entschlossenheit seinen Le bensraum" (Urmersbach 2018, 91–92).

98. "natürliche Aufbau des Waldes" and "Die oberste Schicht wird von weni-gen gebildet. Sie lassen sich zählen. Je weiter hinab, desto mehr Lebensgenossen sind in einer Schicht vereint" (Zechner 2011, 233).

99. Linse 1993, 57.

100. Springer and Sonjevski-Jamrowski 1936.

101. "immerwährende Schicksalsgemeinschaft zwischen deutschem Wald und deutschem Volk" (Zechner 2016, 190).

102. "Aus dem Wald kommen wir / Wie der Wald leben wir / Aus dem Wald formen wir / Heimat und Raum" and "daß weder im Film noch in den Ankündi-gungen" (Linse 1993, 73).

103. Linse 1993, 73.

104. Weinberg 2008, 544, 90 ("fernab von allen menschlichen Siedlungen" and "lauter Kretins" [Jochmann 1980, 351, 104]) (translator's note: a more faithful translation of the German would be "far away from any human settlements" and "a bunch of cretins").

105. Weinberg 2008, 22 ("Von der Schlacht im Teutoburger Wald machen wir uns falsche Vorstellungen. Schuld daran ist die Romantik unserer Geschichtspro-fessoren. Im Wald konnte man damals so wenig wie heute Kämpfe führen" [Joch-mann 1980, 41]) (translator's note: the misspelling of the forest's name is repeated in editions of the English translation).

106. "unterlegene Völker" (Rubner 1982, 122).

107. "der Aufgabe der Holzerzeugung" and "urwüchsige Bestände als Zuflucht hochwertigen Wildes erhalten und gepflegt werden sollen" (Gautschi 2010, 74).

108. Rubner 1997, 184, and Gautschi 2010, 78.

109. "ihrer deutschen Seele entsprechende Umgebung" (Franke 2010, 51); Passeick 2018, 30–31.

110. "Wiederbewaldung des Ostens" (Zechner 2011, 235).

111. "deutsch Waldvolk," "slawisch Steppenvolk" and "jüdisch Wüstenvolk" (Zechner 2016, 191).

112. Gautschi 2010, 223.

113. Kube 1986, 105.

114. Świętorczecki 1938, 9; Gautschi 2010, 96–97 (according to Gautschi, the unedited original can be found at the Institut für Jagdkunde, Göttingen).

115. "germanischen Jagdurwald" (Bode and Emmert 2000, 302).

116. "befrieden" and "evakuieren" (Gautschi 2010, 223–24); Blood 2010, 251.

117. Neumärker and Knopf 2008, 129.

118. "Für Rominten hätte ich mich dem leibhaftigen Teufel verschrieben!" (Frevert 2007, 133).

119. Blood 2010, 250; Rubner 1982, 137.

120. Blood 2010, 251; Neumärker and Knopf 2008, 129.

121. Blood 2010, 251; Gautschi 2010, 224–25.

122. Rubner 1982, 136.

123. "Leider sind aber immer noch Partisanen und sonstige Banditen in großer Zahl hier, und die Strecke an diesen ist ganz erheblich größer als an allem Wild" (Bode and Emmert 2000, 154).

124. Frevert 1977, 220–21; Neumärker and Knopf 2008, 118–19. Nazi propaganda referred to the practice of bombing cities as "coventrieren" ("coventryize") after the Luftwaffe had bombed Coventry in November 1940. According to Victor Klemperer, the term derived by analogy from "magdeburgisieren," which meant "to level to the ground" in the way the town of Magdeburg had been destroyed in May 1631, during the Thirty Years' War (Klemperer 2006, 118–19; cf. Klemperer 1947, 134–35).

125. Frevert 2007, 174.

126. Frevert 1977, 78–81.

127. "sollte auch nur ein einziges feindliches Flugzeug die deutsche Grenze überfliegen" (Kube 1986, 341; Kube also offers a critical analysis of the anecdote).

128. Knopf and Martens 2004, 80–81.

129. "Schön, lassen Sie uns etwas jagen!" (Semmler 1947, 182).

130. Rubner 1982, 172–78; Gautschi 2010, 119.

131. Rubner 1982, 175.

132. Fest 1970, 79–80 (cf. Fest 1963, 114, 435).

133. "Meine armen Hirsche. Das ist entsetzlich!" (Eberle and Uhl 2005, 289).

134. Frevert 1977, 85.

135. Knopf and Martens 2004, 158.

136. Goebbels 1979, 225 ("so ungefähr den Höhepunkt der moralischen Verwirrung Goerings und seiner Umgebung dar" and "an jene bourbonische Prinzessin,

die, als die Masse mit dem Ruf 'Brot!' die Tuilerien stürmte, die naive Frage stellte: 'Warum essen die Leute denn keinen Kuchen?'" [Goebbels 1977, 326]).

137. Perschke 1999, 483–84; Steinbach 2010, 100–101.

138. Kelley 1961, 58.

139. See *Bundesgesetzblatt* 1952. In its hunting legislation from 1953, however, the GDR avoided this fraught term "Waidgerechtigkeit"; see Bode 2016, 67.

Chapter 6. Not Really Stroganoff

1. Piekalkiewicz 1992, 296.

2. On coldbloods, see Zieger 1973, 565–67; on Haflinger, see Sambraus 1991, 112.

3. Buchner 1998, 136.

4. Roscher 2018, 79–80.

5. Schäffer and König, 2015, 1244.

6. von Maltzan 2009, 168–69.

7. Buchner 1998, 137.

8. Meyer 1982, 192; Zieger 1973, 326.

9. Kuhnert 1993, 107.

10. Piekalkiewicz 1992, 296.

11. Krüger 1939, 49–50.

12. Piekalkiewicz 1992, 296.

13. For a description of this river crossing, see Kuhnert 1993, 71–73.

14. Piekalkiewicz 1992, 200.

15. Piekalkiewicz 1992, 98.

16. Piekalkiewicz 1992, 6.

17. Sax 2017, 83.

18. For "verabscheut" ("detested"), see Raulff 2015, 267; see also Raulff 2018, iii. For "dumm" ("dumb"), see Gun 1968, 168.

19. Weinberg 2008, 5 ("im Nu alle Erziehung von sich" [Jochmann 1980, 26]).

20. On Hitler as an *Autonarr* (automobile nut), see Weinberg 2008, 152 ("My weakness is for motor-cars"; literally, "My love belongs to the automobile") ("Meine Liebe gehört dem Automobil" [Jochmann 1980, 161]). For Hitler's view of the cavalry, see Piekalkiewicz 1992, 191.

21. "Kriegskamerad" (Pöppinghege 2009, 237–38).

22. "ritterliche Ideal des Kriegers zu Pferde" (Hürter 2007, 47).

23. "kentaurischen Pakt" (Raulff 2015, 17); see also Raulff 2018.

24. "Ikone des Soldatischen" (Pöppinghege 2009, 238).

25. Diedrich 2008, 44, 95.

26. Piekalkiewicz 1992, 200.

27. Piekalkiewicz 1992, 261.

28. Gewinner 2017, 76.

29. R.-D. Müller 2005, 84.

30. Weinberg 2008, 154 ("Es funktioniert jetzt nur das Panje-Pferd" [Jochmann 1980, 163]).

31. Holl 1978, 48; Clement 2005.

32. Zieger 1973, 573–74.

33. Kuhnert 1993, 107.

34. Meyer 1982, 192.

35. Piekalkiewicz 1992, 50.

36. Kuhnert 1993, 132–33.

37. Meyer 1982, 193.

38. Piekalkiewicz 1992, 50.

39. Zieger 1973, 463.

40. Zieger 1973, 334; Döpke 2004, 29.

41. Zieger 1973, 375; Döpke 2004, 29; Kutter 2012, 145.

42. Buchner 1998, 139.

43. Zieger 1973, 229–30; Hartmann and Lange 2017.

44. Kuhnert 1993, 130.

45. Döpke 2004, 9; on the *Nasenbremse* (twitch), see Jahrbeck 2010, 6.

46. Kuhnert 1993, 149–50).

47. "Zerrissen von Granaten, aufgetrieben, die Augen aus leeren roten Höhlen herausgekugelt, stehend und zitternd, aus einem kleinen Loch in der Brust langsam, aber unaufhörlich blutend, auslaufend—so sehen wir sie nun seit Monaten. Fast ist das schlimmer noch als die weggerissenen Gesichter der Menschen, die verbrannten, halbverkohlten Leichen mit den blutig aufgebrochenen Brustkörben, als die schmalen Blutstreifen hinter dem Ohr der aufs Gesicht Hingebrochenen" (Bähr and Bähr 1952, 87).

48. "Dann hat sich mit faulendem Dachstroh ernährt / Und weitergehungert Kamerad Pferd, / Verwundet, erfroren, von Feuer versehrt / Behalten im Herzen wir Kamerad Pferd" (Döpke 2004, 21).

49. "auf die Klagenden selbst zurück" (De Kleijn 2017, 52).

50. Hürter 2007, 385.

51. "Solange wir noch Pferde haben, geht es, und außerdem wird uns der Führer nicht verlassen" (Ebert 2003, 177).

52. Adam 1965, 220–21.

53. "im Geiste mit" (Wohlfromm and Wohlfromm 2017, 239).

54. "Das letzte Pferd ist schon lange aufgefressen und keine Ahnung wann die Scheiße ein Ende hatt [*sic*]" (Ebert 2003, 290).

55. Ebert 2003, 261.

56. On horse meat, see Holl 1965, 16, and Schaaf 2011.

57. Diedrich 2008, 15.

58. Meyer 1982, 186.

59. Weinberg 2008, 86 (cf. Jochmann 1980, 124); R.-D. Müller 2005, 194.

60. Longerich 2008, 476.

61. Longerich 2008, 288.

62. Kater 2006, 217.

63. R.-D. Müller 2005, 290.

64. "Weißt du noch, Paprika. Wir verstanden uns vom ersten Tag an. Du warst klug, du hattest Temperament. Auf den leisesten Schenkeldruck hast du reagiert. Natürlich war dein Trab fürchterlich. Warum mußtest du auch die Vorderbeine so

hoch werfen? Sei mir nicht böse, Paprika, oft hatte ich dich im Verdacht, aus der leichtfertigen Zirkuswelt zu stammen. Aber dein herrlicher, dein unvergleichlicher Galopp, deine Schnelligkeit, dein Springvermögen waren famos," "Das Biest schlägt dir noch einmal die Knochen entzwei," "Einmal beißt dir das Biest noch die Nase ab," "Mit gespitzten Ohren und leise schnaubend, hast du dich, unendlich langsam, in der eigenen Spur zurückgesetzt," "Nie habe ich dir gesagt, Paprika, daß ich dich einmal nach einem glücklichen Kriegsende kaufen wollte, daß ich dazu schon für dich eine Unterkunft in Berlin bei guten Menschen gefunden hatte. Nun müssen wir uns trennen, ein unerbittliches Kriegsschicksal reißt uns auseinander," and "Noch einmal zog ich den warmen Atem Paprikas tief ein . . . noch einmal legte ich mein Gesicht an ihre samt weichen Nüstern. Ich sah ihr nach, bis der Wall mir den Blick versperrte" (Piekalkiewicz 1992, 282–83; Meyer 1982, 191).

65. Meyer 1980, 186.

66. Meyer 1982, 191; Sax, 2017, 86.

67. Piekalkiewicz 1992, 296.

68. Meyer 1982, 192.

69. "Mit Pferden ließ er sich nicht gewinnen und ohne Pferde erst recht nicht" (Koselleck 2005, 172).

70. Henseleit 2006, 1:193.

71. Henselei 2006, 1:70.

72. Henseleit 2006, 1:196–97.

73. Henseleit 2006, 1:265.

74. Henseleit 2006, 1:197.

75. For Bleeker's role in National Socialism, see Henseleit 2006, 1:40; on the "Gottbegnadeten-Liste" (list of those graced by God), see Klee 2009, 53.

76. "deutschen Soldatenpferde," "fromm, willig und ausdauernd bis zum letzten Atemzug," and "geistigen Gehalt" (Henseleit 2006, 1:198); see also Henseleit 2006, 1:198, on the cavalry memorial in general.

77. Canetti 1978, 153 ("Das schönste Standbild des Menschen wäre ein Pferd, wenn es ihn abgeworfen hätte" [Canetti 1993, 203]).

78. Personal correspondence with the Bundesverteidigungsministerium (German Defense Ministry), January 24, 2017, and March 31, 2017.

79. "Fleischskandal" 2013.

80. "Die fehlende Deklaration, also der Betrug am Verbraucher, ist der eigentliche Skandal" (Lüdemann 2013).

81. Harris 1995, 93–104.

82. Sambraus 1991, 105–6.

83. "Kleinmädchenhysterie" and "Nirgendwo ist der Ekel vor Pferdefleisch so ausgeprägt wie in der Generation, die noch den Zweiten Weltkrieg und die Nachkriegsjahre erlebt haben" (Heine 2013).

84. "eine unangenehme Erinnerung, vor allen Dingen bei älteren Leuten" and "Notschlachtungen von Pferden, an Stalingrad Bilder" (Fischer 2013).

85. "die aus Scham nicht betrauert werden" (Pöppinghege 2009, 240).

Epilogue

1. Speer 2008, 302 ("Ich werde einmal nur zwei Freunde haben, . . . Fräulein Braun und meinen Hund," "Tapferkeit seiner Mätresse" and "Abhängigkeit seines Hundes" [Speer 1969, 314]).

2. "Seit ich die Menschen kenne, liebe ich die Hunde" (Zoller 1949, 230).

3. Gun 1968, 168.

4. Eberle and Uhl 2007, 291.

5. Junge 2011, 153.

6. Zoller 1949, 151.

7. Piekalkiewicz 1992, 297; Trakehner Verband 2021.

8. "Deutschlands klassisches Pferdereservoir" and "russisch besetzt" ("Keine Staatsamateure mehr" 1951, 22).

9. Schlange-Schoeningen 1948, 232 ("Alle Wagen werden fertiggemacht. Die meisten Leute sind in kopfloser Angst. Ich werde dafür sor gen, daß sie so geordnet wie möglich abtransportiert werden. Meine tapfere Frau und ich bleiben. Hitlers Bonzen fliehen," and "in grimmiger Kälte, bei Eis und Schnee und Sturm. . . . Tausende von verendeten Pferden an den Wegrändern. Tote Menschen im Schnee verscharrt. Nur vorwärts, vorwärts: Die Russen kom men! Napoleons Rückzug muß ein Kinderspiel gewesen sein" [H. Schlange-Schöningen 1946, 202–3]).

10. Muth 1970, 412.

11. Schlange-Schoeningen 1948, 235 ("Schöningen war ein brennender Trümmerhaufen. Ich bin in der Fremde" [H. Schlange-Schöningen 1946, 204]).

12. Falkenberg and Hammer 2008, 317.

13. Sächsische Landesanstalt für Landwirtschaft 2003, 30.

14. "In langen unabsehbaren Schlangen klappern die Wagen über das zerrissene Pflaster der Ruinenstadt. . . . Das Land kam in die Stadt; wir sind um vieles der Natur näher gerückt. Unser ganzes Leben ist bis in die Wirtschaft hinein vom Pferd abhängig" ("Panjepferde" 1945, 3).

15. "Ihr glänzendes Fell ist struppig geworden, und hervorstehende Rippen und Hüfthöcker zeigen nur zu deutlich, daß ihnen oft der Magen knurrt. . . . Da klappern die Hufeisen oder fehlen gar ganz, schlechtsitzende Geschirre scheuern oft handtellergroße Wunden am Kopf, am Kamm oder am Bauch des Tieres" and "Vielleicht haben wir uns schon zu sehr an die maschinelle Gefühllosigkeit der Motoren gewöhnt, sonst würde wohl mancher Wagenbesitzer einen liebevolleren Blick für das schwere Leben seines Zugpferdes besitzen" ("Pferdelenker oder Pferdeschinder?" 1945, 3).

16. "war nicht in Stalingrad!" (Beyer 1987, 29).

17. Humburg 1998, 147–48, 152.

18. "Wanzen in der Klei dung, Läuse in der Wäsche und die Flöhe zwischendurch! Xmal zieht man sich aus, sucht alles ab, nachdem ist es dasselbe," "Wenn es möglich ist irgendeine Salbe oder dergl. zu erhalten, die man als Abwehrmittel gebrauchen kann schickt bitte so etwas," and "Unseren Läusen geht es sehr gut; sie vermehren sich am laufenden Band" (Humburg 1998, 151–53).

19. Levi 1985, 91 ("mögen nicht sehr sympathische Tierchen sein, aber sie haben keine rassistischen Vorurteile" [Levi 1994, 76]).

20. Vinogradov, Pogonyi, and Teptzov 2005, 88.

21. Junge 2011, 141.

22. "von den Russen in einem Panoptikum" (Eberle and Uhl 2007, 408); Fest 1974, 747 ("Exponat im Moskauer Zoo" [Fest 2013, 1052]).

23. Fest 2004b, 105 (cf. Fest 2004a, 126); Speer 2008, 479 (cf. Speer 1969, 482).

24. On the killing of Blondi, see, among others, Vinogradov, Pogonyi, and Teptzov 2005, 194, Eberle and Uhl 2007, 441, Fest 1974, 748 (cf. Fest 2013, 1055), Junge 2011, 200, and Misch 2010, 218.

25. Eberle and Uhl 2007, 451.

26. On the circumstances of Hitler's death, see Fest 2004b, 115–16 (cf. Fest 2004a, 137–38).

27. "Immer mit Dir" (Klimenko 1965).

28. "Ich glaube, ein Mensch, der gegen ein treues Tier gleichgültig sein kann, wird gegen seinesgleichen nicht dankbarer sein, und wenn man vor die Wahl gestellt wird, ist es besser, zu empfindsam, als zu hart zu sein" (von Preußen and von Preußen 2012, 12).

29. "scheint das Bild des Dritten Reiches nicht perfekt zu sein" (Wippermann and Berentzen 1999, 86).

30. "in der der Mensch dem Menschen zum Wolf [wurde]" (Jähner 2019, 10).

31. On the "Würger," see "Wo der Hund begraben liegt" 1948 and Wiborg 2007.

32. "und die erfundenen Fabelwesen verschwanden aus der erhitzten Fantasie der Bevölkerung. Im Lichtenmoor grast wieder das Vieh in beschaulichem Frieden" (Bundesarchiv 1948).

WORKS CITED

Note: In rendering this work from the German, the translator has drawn on extant English translations, as noted among the following sources, when accessible. Otherwise, all translations from German into English are his own.

Ackermann, Josef. 1993. "Heinrich Himmler: 'Reichsführer-SS.'" In *The Nazi Elite*, edited by Ronald Smelser and Rainer Zitelmann, translated by Mary Fischer, 98–112. New York: New York University Press.

Adam, Wilhelm. 1965. *Der schwere Entschluß*. Berlin: Verlag der Nation.

Adorno, Theodor W. 1951. *Minima Moralia: Reflexionen aus dem beschädigten Leben*. Berlin: Suhrkamp.

Adorno, Theodor W. 1974. *Minima Moralia: Reflections from Damaged Life*. Translated by E. F. N. Jephcott. London: Verso.

Ahlborn, Eduard, dir. 1937. *Internationale Jagdausstellung in Berlin 1937*. Agentur Karl Höffkes: Filmarchiv. http://archiv-akh.de/filme?utf-8=✓&q=jagdausstellung#1.

Ahne, Petra. 2016. *Wölfe*. Berlin: Matthes & Seitz.

Alcabes, Philip. 2009. *Dread: How Fear and Fantasy Have Fueled Epidemics from the Black Death to Avian Flu*. New York: PublicAffairs.

Aly, Götz. 2005. *Hitlers Volksstaat: Raub, Rassenkrieg und nationaler Sozialismus*. Frankfurt am Main: Fischer.

Ammon, Otto. 1906. *Die Bedeutung des Bauernstandes für den Staat und die Gesellschaft*. Berlin: Trowitsch.

Anonymous, dir. 1936. *Goering bedankt sich beim Volk nach Autounfall*. https://archive.org/details/1936-Goering-bedankt-sich-beim-Volk.

Arnold, Dietmar, and Reiner Janick. 2005. *Neue Reichskanzlei und "Führerbunker": Legenden und Wirklichkeit*. Berlin: Links.

Backe, Herbert. 1943. "Aufrechterhaltung der Schweinebestände durch Erweiterung der Futtergrundlage über den Hackfruchtbau." *Zeitschrift für Schweinezucht* 19:1.

Bähr, Walter, and Hans Walter Bähr. 1952. *Kriegsbriefe gefallener Studenten: 1939–1945*. Tübingen: Wunderlich.

Barth, Boris. 2014. "Tiere und Rasse: Menschenzucht und Eugenik." In *Tiere und Geschichte: Konturen einer "Animate History,"* edited by Gesine Krüger, Aline Steinbrecher, and Clemens Wishermann, 199–217. Stuttgart: Steiner.

Bäumer-Schleinkofer, Änne. 1992. "Biologie unter dem Hakenkreuz: Biologie und Schule im Dritten Reich." *Universitas: Zeitschrift für interdisziplinäre Wissenschaft* 47, no. 1: 48–61.

Bayerische Staatsbibliothek. 1943. "Rede des Reichsführers SS bei der SS-Gruppenführertagung in Posen am 4. Oktober 1943." 100(0) Schlüsseldokumente zur deutschen Geschichte im 20. Jahrhundert. https://www.1000dokumente.de/index.html?c=dokument_de&dokument=0008_pos&object=pdf&st=&l=de.

Bein, Alexander. 1965. "Der jüdische Parasit." *Vierteljahreshefte für Zeitgeschichte* 13, no. 2: 121–49.

Benz, Wolfgang. 2011. "Die Schatten der Vergangenheit: Richard Glazar." In *Deutsche Juden im 20. Jahrhundert: Eine Geschichte in Porträts*, edited by Wolfgang Benz, 299–301. Munich: Beck.

Benz, Wolfgang, and Barbara Distel, eds. 2005. *Die Organisation des Terrors.* Vol. 1 of *Der Ort des Terrors: Geschichte der nationalsozialistischen Konzentrationslager.* Munich: Beck.

Berg, Anne. 2015. "The Nazi Rag-Pickers and Their Wine: The Politics of Waste and Recycling in Nazi Germany." *Social History* 40, no. 4: 446–72.

Bergander, Götz. 1977. *Dresden im Luftkrieg.* Cologne: Böhlau.

Beyer, Wilhelm Raimund. 1987. *Stalingrad: Unten, wo das Leben konkret war.* Frankfurt am Main: Athenäum.

Bieger, Walter, ed. 1940. *Handbuch der Deutschen Jagd.* Berlin: Parey.

Blood, Philip W. 2010. "Securing Hitler's Lebensraum: The Luftwaffe and Białowieża Forest, 1942–1944." *Holocaust and Genocide Studies* 24, no. 2: 247–72.

Bluhm, Detlef. 2013. *Was Sie schon immer über Katzen wissen wollten.* Berlin: Insel.

Blumesberger, Susanne. 2009. "Von Giftpilzen, Trödeljakobs und Kartoffelkäfern—Antisemitische Hetze in Kinderbüchern während des Nationalsozialismus." *Medaon—Magazin für jüdisches Leben in Forschung und Bildung* 5, no. 3: 1–13.

Boberach, Heinz. 1982. *Jugend unter Hitler.* Düsseldorf: Droste.

Bode, Wilhelm. 2016. "'Die anerkannten Grundsätze der deutschen Weidgerechtigkeit' gem. § 1 Abs. 3 BjagdG—ein trojanisches Pferd der völkischen Rechtserneuerung im Jagdrecht?" *Jahrbuch des Agrarrechts* 13:33–121.

Bode, Wilhelm. 2018. *Hirsche: Ein Porträt.* Berlin: Matthes & Seitz.

Bode, Wilhelm, and Elisabeth Emmert. 2000. *Jagdwende: Vom Edelhobby zum ökologischen Handwerk.* Munich: Beck.

Bornemann, Gundula. 1953. *50 Jahre Deutsche Edelschweinzucht.* Radebeul: Neumann.

Brehm, Alfred Edmund. 1876. Säugethiere. Vol. 1, pt. 1, of *Brehms Tierleben: Allgemeine Kunde des Tierreichs.* Säugethiere. Leipzig: Verlag des Bibliographischen Instituts.

Brehm, Alfred Edmund. 1877. *Wirbellose Thiere.* Vol. 1, pt. 4, of *Brehms Thierleben: Allgemeine Kunde des Thierreichs.* Leipzig: Verlag des Bibliographischen Instituts.

Brucker, Renate. 2015. "Für eine radikale Ethik: Die Tierrechtsbewegung in der ersten Hälfte des 20. Jahrhunderts." In *Das Mensch-Tier-Verhältnis: Eine sozialwissenschaftliche Einführung*, edited by Renate Brucker, Melanie Bujok, Birgit Mütherich, Martin Seeliger, and Frank Thieme, 211–68. Wiesbaden: Springer.

Brümmer, Johannes. 1877. "Ein überschätzter Feind." *Die Gartenlaube* 31:522–25.

Buchheim, Christoph. 2010. "Der Mythos vom 'Wohlleben': Der Lebenswandel der deutschen Zivilbevölkerung im Zweiten Weltkrieg." *Vierteljahresschrift für Zeitgeschichte* 58, no. 3: 299–328.

Buchner, Leander. 1998. "Zur Bedeutung des Pferdes in der Wehrmacht." In *Veterinärmedizin im Dritten Reich*, 135–44. Gießen: Deutsche Veterinärmedizinische Gesellschaft.

Bundesarchiv. 1948. "Der Würger vom Lichtenmoor." *Welt im Film*, episode 171. https://www.filmothek.bundesarchiv.de/video/583603.

Bundesarchiv. 2019. "Persecution and Murder of the Jewish Population in Germany, 1933–1945." *Victims of the Persecution of Jews under the National Socialist Tyranny in Germany, 1933–1945.* https://www.bundesarchiv.de/gedenkbuch/introduction/#persecution.

Bundesgesetzblatt. 1952. "Bundesjagdgesetz" (Federal Hunting Act). Pt. 1, 780–88.

Burchardt, Lothar. 2007. "Von Katzen und Mäusen: Einige Bemerkungen zur Denominationskultur der deutschen Streitkräfte im 20. Jahrhundert." In *Von Katzen und Menschen: Sozialgeschichte auf leisen Sohlen*, edited by Clemens Wischermann, 183–210. Konstanz: UVK Verlagsgesellschaft.

Canetti, Elias. 1978. *The Human Province.* Translated by Joachim Neugroschel. New York: Seabury Press.

Canetti, Elias. 1984. *Crowds and Power.* Translated by Carol Stewart. New York: Farrar, Straus and Giroux.

Canetti, Elias. 1993. *Die Provinz des Menschen.* In *Aufzeichnungen 1942–1985.* Vol. 4 of *Gesammelte Werke.* Munich: Hanser.

Canetti, Elias. 1994. *Masse und Macht.* Vol. 3 of *Gesammelte Werke.* Munich: Hanser.

Clement, Richard M., Jr. 2005. "Interview with Oswald Maier." *Der erste Zug.* http://www.dererstezug.com/VetMaier.htm.

Corni, Gustavo. 1993. "Richard Walther Darré: The Blood and Soil Ideologue." In *The Nazi Elite*, edited by Ronald Smelser and Rainer Zitelmann, translated by Mary Fischer, 18–27. New York: New York University Press.

Corni, Gustavo, and Horst Gies. 1997. *Brot—Butter—Kanonen: Die Ernährungswirtschaft in Deutschland unter der Diktatur Hitlers.* Berlin: Akademie.

Darré, Richard Walther. 1933. *Das Schwein als Kriterium für nordische Völker und Semiten.* Munich: Lehmann.

Darré, Richard Walther. 1937. *Der Schweinemord.* Munich: Eher.

Darwin, Charles. 1868. *The Variation of Animals and Plants under Domestication.* Vol. 2. London: Murray.

de Kleijn, David M. 2017. "Von Reitergeist und stummen Kameraden: Narrative Leitsemantiken des Pferdebildes im deutschen Weltkriegsgedenken." *Tierstudien* 6, no. 12: 46–56.

Demandt, Alexander. 2007. *Das Privatleben der römischen Kaiser.* Munich: Beck.

Denk, Manfred. 2011. "Die Konstruktion der jüdischen 'Rasse': Ein Ideologievergleich der Rasse-Konzepte H. S. Chamberlains und A. Hitlers, durchgeführt an ihren Hauptwerken *Grundlagen des 19. Jahrhunderts* bzw. *Mein Kampf.*" PhD diss., University of Erlangen-Nuremberg.

Deutsche Wochenschau. 1941. Newsreel no. 62783, reel 3. https://www.net-film
.ru/en/film-62783.

Deutscher Jagdverband. 2016. "Hubertustag—Warum wir Jäger diesen Tag in
Ehren halten." *Jäger—Zeitschrift für das Jagdrevier*, November 3. https://www
.jaegermagazin.de/jagd-aktuell/news-fuer-jaeger/hubertustag-warum-wir-jaeger
-diesen-tag-in-ehren-halten.

Dieberger, Johannes. 2018. "Geschichte der Jagdkultur und das Dritte Reich: Auswirkungen auf unser heutiges Weidwerk, Teil II." *Der OÖ Jäger: Info-Magazin
des OÖ Landesjagdverbandes* 45, no. 159: 68–73.

Diedrich, Torsten. 2008. *Paulus: Das Trauma von Stalingrad: Eine Biographie.* Paderborn: Schöningh.

Diels, Rudolf. 1949. *Lucifer ante portas: Zwischen Severing und Heydrich.* Zürich:
Interverlag.

Dinzelbacher, Peter, ed. 2000. *Mensch und Tier in der Geschichte Europas.* Stuttgart:
Kröner.

Dirscherl, Stefan. 2012. *Tier- und Naturschutz im Nationalsozialismus: Gesetzge-
bung, Ideologie und Praxis.* Göttingen: V&R Unipress.

Döpke, Oswald. 2004. *Ich war Kamerad Pferd: Meine grotesken Kriegserlebnisse,
1942–1945.* Berlin: Zeitgut.

Dornheim, Andreas. 2011. *Rasse, Raum und Autarkie: Sachverständigengutachten
zur Rolle des Reichsministeriums für Ernährung und Landwirtschaft in der NS-
Zeit.* Bonn: Bundesministerium für Ernährung, Landwirtschaft und Verbraucher-
schutz. https://www.bmel.de/SharedDocs/Downloads/DE/_Ministerium/Ges
chichte/sachverstaendigenrat-zur-rolle-ns-zeit.html.

Eberle, Henrik, and Matthias Uhl, eds. 2007. *Das Buch Hitler: Geheimdossier des
NKWD für Josef W. Stalin, zusammengestellt aufgrund der Verhörprotokolle des
Persönlichen Adjutanten Hitlers, Otto Günsche, und des Kammerdieners Heinz
Linge, Moskau 1948/49.* Bergisch Gladbach: Bastei Lübbe.

Ebert, Jens, ed. 2003. *Feldpostbriefe aus Stalingrad: November 1942 bis Januar 1943.*
Göttingen: Wallstein.

Editors. 1941. Foreword to *Zeitschrift für Schweinezucht*, no. 10: 1.

"Ein erschütterndes Filmdokument." 1938. *Kleines Volksblatt*, August 28.

Falkenberg, Heinz, and Horst Hammer. 2006. "Mitteilung: Zur Domestikation
und Verbreitung der Hausschweine in der Welt." Pt. 1 of "Zur Geschichte und
Kultur der Schweinezüchtung und-haltung." *Züchtungskunde* 78, no. 1: 55–68.

Falkenberg, Heinz, and Horst Hammer. 2008." Mitteilung: Schweinezucht und-
produktion in Europa zwischen 1900 und 1945." Pt. 4 of "Zur Geschichte und
Kultur der Schweinezüchtung und-haltung." *Züchtungskunde* 80, no. 4: 315–33.

Fehringer, Otto. 1953. *Die Welt der Säugetiere.* Munich: Droemer.

Fest, Joachim C. 1963. *Das Gesicht des Dritten Reiches: Profile einer totalitären
Herrschaft.* Munich: Piper.

Fest, Joachim C. 1970. *The Face of the Third Reich: Portraits of the Nazi Leadership.*
Translated by Michael Bullock. London: Weidenfeld & Nicholson.

Fest, Joachim C. 1974. *Hitler.* Translated by Richard and Clara Winston. New
York: Harcourt, Brace.

Fest, Joachim C. 2004a. *Der Untergang: Hitler und das Ende des Dritten Reiches, eine historische Skizze.* Reinbek: Rowohlt Taschenbuch.

Fest, Joachim C. 2004b. *Inside Hitler's Bunker: The Last Days of the Third Reich.* Translated by Margot Bettauer Dembo. New York: Picador.

Fest, Joachim C. 2013. *Hitler: Eine Biographie.* Frankfurt am Main: Ullstein.

Fischer, Karin. 2013. "Furcht vor dem Pferd: Kulturgeschichte eines Verzehrverbots." *Deutschalndfunk*, February 14. http://www.deutschlandfunk.de/furcht-vor -dem-pferd.691.de.html?dram:article_id=237597.

"Fleischskandal: Pferdemedikament gelangte ins Essen." 2013. *Spiegel Online*, February 14. http://www.spiegel.de/wirtschaft/soziales/pferde- medikament-soll-ins -essen-gelangt-sein-a-883372.html.

Flessau, Kurt-Ingo. 1977. *Schule der Diktatur: Lehrpläne und Schulbücher des Nationalsozialismus.* Munich: Ehrenwirth.

Föger, Benedikt, and Klaus Taschwer. 2001. *Die andere Seite des Spiegels: Konrad Lorenz und der Nationalsozialismus.* Vienna: Czernin.

Fontane, Theodor. 1910. *Wanderungen durch die Mark Brandenburg.* Stuttgart: Cotta.

Franke, Nils M. 2010. "Naturschutz und Rechtsextremismus: Historische und aktuelle Befunde." *Denkanstöße: Naturschutz und Wissenschaft*, April 8, 51–53.

Freund, Florian. 1989. *Arbeitslager Zement: Das Konzentrationslager Ebensee und die Raketenrüstung.* Vienna: Verlag für Gesellschaftskritik.

Frevert, Walter. 1977. *Rominten.* Munich: BLV-Verlagsgesellschaft.

Frevert, Walter. 2007. *Mein Jägerleben: Gesammelte Erzählungen des großen Waidmanns.* Stuttgart: Kosmos.

Freytag, Otto. 2013. "Teenager an der Flugabwehrkanone." Interview with Zeitzeugen-Portal, Stiftung Haus der Geschichte der Bundesrepublik Deutschland, November 7. https://www.youtube.com/watch?v=GmfuMA5WaQk.

Fröhlich, Elke. 2013. *Der Zweite Weltkrieg: Eine kurze Geschichte.* Stuttgart: Reclam.

Fuhr, Eckhard. 2016. *Rückkehr der Wölfe: Wie ein Heimkehrer unser Leben verändert.* Munich: Goldmann.

Gautschi, Andreas. 2010. *Der Reichsjägermeister: Fakten und Legenden um Hermann Göring.* Melsungen: Nimrod.

Geißler, Erhard. 1998. *Hitler und die Biowaffen.* Münster: Lit.

Geißler, Erhard. 1999. *Biologische Waffen—nicht in Hitlers Arsenalen: Biologische und Toxin-Kampfmittel in Deutschland von 1915 bis 1945.* Münster: Lit.

Gerhard, Gesine. 2012. "Das Bild der Bauern in der modernen Industriegesellschaft: Störenfriede oder Schoßkinder der Industriegesellschaft?" In *Das Bild des Bauern. Selbst- und Fremdwahrnehmungen vom Mittelalter bis ins 21.Jahrhundert*, edited by Daniela Münkel and Frank Uekötter, 111–30. Göttingen: Vandenhoeck & Ruprecht.

Gerhard, Gesine. 2015. *Nazi Hunger Politics: A History of Food in the Third Reich.* London: Rowman & Littlefield.

Gerriets, Jan. 1933. "Mai!" *Zeitschrift für Schweinezucht*, no. 21: 329–31.

Gewinner, Malin. 2017. *Die Anthropomorpha: Tiere im Krieg.* Berlin: Matthes & Seitz.

Gies, Horst. 2019. *Richard Walther Darré: Der "Reichsbauernführer," die national-sozialistische "Blut und Boden"-Ideologie und Hitlers Machteroberung.* Cologne: Böhlau.

Giese, Clemens, and Waldemar Kahler. 1944. *Das deutsche Tierschutzrecht: Bestimmungen zum Schutze der Tiere.* Berlin: Duncker & Humblot.

Glazar, Richard. 1983. "Wie schwer wiegt das 'Nichts'?" *Die Zeit,* October 21, 71–72.

Glazar, Richard. 1996. Interview with the Visual History Archive, USC Shoah Foundation, February 5. Transcribed by the Freie Universität Berlin. http://transcripts.vha.fu-berlin.de/interviews/40?locale=de&query=glazar.

Goebbels, Joseph. 1935. "Was wollen wir im Reichstag?" In *Der Angriff: Aufsätze aus der Kampfzeit,* edited by Hans Schwarz van Berk, 73. Munich: Eher.

Goebbels, Joseph. 1948. *The Goebbels Diaries, 1942–1943.* Edited and translated by Louis L. Locher. New York: Doubleday.

Goebbels, Joseph. 1977. *Tagebücher 1945: Die letzten Aufzeichnungen.* Hamburg: Hoffmann u. Campe.

Goebbels, Joseph. 1979. *Final Entries, 1945: The Diaries of Joseph Goebbels.* Edited by Hugh Trevor-Roper. Translated by Richard Barry. New York: Avon.

Goebbels, Joseph. 1992. *Tagebücher, 1924–45.* Vol. 3, *1935–39.* Munich: Piper.

Goebbels, Joseph. 2005. *Die Tagebücher von Joseph Goebbels,1923–1941.* Vol. 3, pt. 1, *April 1934–Februar 1936.* Munich: De Gruyter.

Göring, Emmy. 1967. *An der Seite meines Mannes: Begebenheiten und Bekenntnisse.* Göttingen: Schütz.

Göring, Hermann. 1938. Preface to *Waidwerk der Welt.* Edited by the Reichsbund Deutsche Jägerschaft. Berlin: Parey.

Grimm, Jacob, and Wilhelm Grimm. 1884. *Grimm's Household Tales.* Translated and edited by Margaret Hunt. London: George Bell.

Grimm, Jacob, and Wilhelm Grimm. 1981. *Kinder- and Hausmärchen.* Vol. 2. Berlin: Insel.

Gritzbach, Erich. 1938. *Hermann Göring: Werk und Mensch.* Munich: Zentralverlag der NSDAP.

Grün, Max von der. 1979. *Wie war das eigentlich? Kindheit und Jugend im Dritten Reich.* Darmstadt: Luchterhand.

Gun, Nerin E. 1968. *Eva Braun-Hitler: Leben und Schicksal.* Velbert: Blick + Bild.

Hackett, David A., ed. 1995. *The Buchenwald Report.* Boulder, CO: Westview Press.

Haeckel, Ernst. 1904. *Die Lebenswunder: Gemeinverständliche Studien über biologische Philosophie.* Stuttgart: Kröner.

Haikal, Mustafa, and Jörg Junhold. 2003. *Auf der Spur des Löwen: 125 Jahre Zoo Leipzig.* Leipzig: Pro Leipzig.

Harris, Marvin, 1995. *Wohlgeschmack und Widerwillen: Die Rätsel der Nahrungstabus.* Munich: Deutscher Taschenbuch.

Harris, Marvin. 1998. *Good to Eat: Riddles of Food and Culture.* Long Grove, IL: Waveland.

Hartmann, Volker, and Heike Lange. 2017. "'. . . Ausbilden, Erziehen und Führen': Interview mit dem leitenden Veterinär der Bundeswehr, Oberstveterinär Dr.

Leander Buchner." *Wehrmedizin und Wehrpharmazie*, October 1. http://www
.wehrmed.de/article/3273-ausbilden-erziehen-fuehren.html.

Hasel, Karl. 1985. *Forstgeschichte: Ein Grundriss für Studium und Praxis.* Vol. 1.
Hamburg: Parey.

Haus der Wannsee-Konferenz. 2021. "Dokumente bis zur Konferenz 1942."
Gedenk- und Bildungsstätte. https://www.ghwk.de/wannsee-konferenz/doku
mente-zur-wannsee-konferenz.

Hecht, Heinrich. 1979. "Schweinezucht in Pommern und der Verband pommer-
scher Schweinezüchter." *Pommersche Zeitung*, August 18, 12–13.

Heine, Matthias. 2013. "Der Ekel vor Pferdefleisch kommt aus Stalingrad." *Die Welt*,
February 18. https://www.welt.de/kultur/article113726494/Der-Ekel-vor-Pferde
fleisch-kommt-aus-Stalingrad.html.

Henseleit, Frank. 2006. "Der Bildhauer Bernhard Bleeker (1881–1968): Leben und
Werk." 2 vols. PhD diss., Augsburg.

Herder, Johann Gottfried von. 1800. *Outlines of a Philosophy of the History of Man.*
Translated by T. Churchill. London: Johnson.

Herder, Johann Gottfried von. 1841. *Ideen zur Philosophie der Geschichte der Menschheit.* Vol. 2. Riga: Hartknoch.

Herlemann, Beatrix, 1993. *"Der Bauer klebt am Hergebrachten": Bäuerliche Ver-
haltensweisen unterm Nationalsozialismus auf dem Gebiet des heutigen Landes
Niedersachsen.* Hannover: Hahn.

Hermand, Jost. 1994. *Als Pimpf in Polen: Erweiterte Kinderlandverschickung 1940–
1945.* Frankfurt am Main: Fischer Taschenbuch.

Herrmann, Bernd. 2009. "Kartoffel, Tod und Teufel: Wie Kartoffel, Kartoffelfäule
und Kartoffelkäfer Umweltgeschichte machten." In *Schauplätze und Themen
der Umweltgeschichte*, edited by Bernd Herrmann and Urte Stobbe, 71–120.
Göttingen: Universitätsverlag Göttingen.

Herzig, Arno. 2010. "1933–1945: Vertreibung und Vernichtung." *Jüdisches Leben in
Deutschland: Informationen zur politischen Bildung*, no. 307: 51–61.

Hesse, Hermann, and Thomas Mann. 1999. *Briefwechsel.* Edited by Anni Carlsson
and Volker Michels. Berlin: Suhrkamp.

Hiemer, Ernst. 1938. *Der Giftpilz: Ein Stürmerbuch für Jung und Alt.* Nuremberg:
Der Stürmer.

Hiemer, Ernst. 1940. *Der Pudelmopsdackelpinscher und andere besinnliche Erzählungen.* Nuremberg: Der Stürmer.

Himmler, Heinrich. 1943. "Speech of the *Reichsführer*-SS at the SS Group Leader
Meeting in Posen." October 4. http://www.holocaustresearchproject.org/holo
prelude/posen.html.

Himmler, Heinrich. 1946. "Speech of the Reichsfuehrer-SS Heinrich Himmler at
Kharkow, April 1943." In *Nazi Conspiracy and Aggression*, 4:572–78. Washington,
DC: U.S. Government Printing Office.

Hippler, Fritz, dir. 1940. *Der ewige Jude: Dokumentarfilm über das Weltjudentum.*
Deutsche Filmherstellungs- und Verwertungs-GmbH.

Höge, Helmut. 2017. "Über Tiernamen als Waffen." *Tierstudien* 6, no. 12: 95–107.

Hohlbaum, Robert. 1932. "Von einem Ehepaar und einem Junggesellen." In *Tiere im Krieg*, edited by Johannes Theuerkauff, 186–90. Berlin: Kolk.

Holl, Adelbert. 1965. *Was geschah nach Stalingrad? 7¼ Jahre als Kriegs- und Strafgefangener in Russland.* Erlangen: Müller.

Holl, Adelbert. 1978. *Als Infanterist in Stalingrad: Bericht.* Erlangen: Müller.

Holtz, Tobias. 2018. "Zwischen Idylle und Hölle." *Nordkurier*, March 12.

Hoess, Rudolph. 1951. *Commandant of Auschwitze: The Autobiography of Rudolph Hoess.* Cleveland: World Publishing Company.

Höß, Rudolf. 2006. *Kommandant in Auschwitz: Autobiografische Aufzeichnungen des Rudolf Höß.* Edited by Martin Broszat. Munich: Deutscher Taschenbuch.

Hopp, Paul-Joachim, and Wolfgang Weitz. 2017. "Walter Frevert—Der Macht verfallen." *Wild und Hund* 13.

Höxter, Herta. 1997. Interview by the Visual History Archive, USC Shoah Foundation, July 14. Transcribed by the Freie Universität Berlin. http://transcripts.vha.fu-berlin.de/interviews/623.

Humburg, Martin. 1998. *Das Gesicht des Krieges: Feldpostbriefe von Wehrmachtsoldaten aus der Sowjetunion, 1941–1944.* Wiesbaden: Verlag für Sozialwissenschaften.

Hürter, Johannes. 2007. *Hitlers Heerführer: Die deutschen Oberbefehlshaber im Krieg gegen die Sowjetunion, 1941/42.* Munich: Oldenbourg.

Hutter, Ralf. 2016. "Die Schule in Witzenhausen." *Neues Deutschland*, September 17. https://www.nd-aktuell.de/artikel/1025729.die-schule-in-witzenhausen.html.

Huxley, Aldous. 1922. *Eyeless in Gaza.* London: Chatto & Windus.

Huxley, Aldous. 1987. *Geblendet in Gaza.* Translated by Herberth E. Herlitschka. Munich: R. Piper.

Industrieverband Agrar. 2013. "Kartoffelkäfer—ein Schädling mit Geschichte." *IVA-Magazin Online*, July 19. https://www.iva.de/iva-magazin/schule-wissen/kartoffelkaefer-ein-schaedling-mit-geschichte.

Jäckel, Eberhard. 1992. "Die Konferenz am Wannsee." *Die Zeit*, January 17, 33.

Jähner, Harald. 2019. *Wolfzeit: Deutschland und die Deutschen 1945–1955.* Berlin: Rowohlt Taschenbuch.

Jahrbeck, Andrea. 2010. "Zur Anwendung von Zwangsmaßnahmen bei Pferden unter besonderer Berücksichtigung der Nasenbremse unter Tierschutz." *Merkblatt* 129:6.

Jansen, Sarah. 2003. *"Schädlinge": Geschichte eines wissenschaftlichen und politischen Konstrukts 1840–1920.* Frankfurt am Main: Campus.

Jentzsch, Rupert. 1998. "Tierschutz und Schächtverbot: Das Schlachtgesetz von 1933 und seine Auswirkungen." In *Veterinärmedizin im Dritten Reich*, edited by Johann Schäffler, 44–53. Gießen: Deutsche Veterinärmedizinische Gesellschaft.

Jochmann, Werner, ed. 1980. *Adolf Hitler: Monologe im Führerhauptquartier, 1941–1944.* Munich: Heyne.

Josselin, Blandine, dir. 2017. *Hitler und der Wolf: Rassenwahn im Dritten Reich.* https://www.youtube.com/watch?v=mJvjhNNuLL4.

Junge, Traudl. 2011. *Hitler's Last Secretary: A Firsthand Account of Life with Hitler.* Translated by Melissa Müller. New York: Skyhorse.

Kalm, Ernst. 1996. "Schweinezucht in Pommern—vom Fett- zum Fleischschwein." In *Tierzucht in Pommern*, 71–93. Kiel: Stiftung Pommern.

Kater, Michael H. 2006. *Das "Ahnenerbe" der SS 1935–1945: Ein Beitrag zur Kulturpolitik des Dritten Reiches*. Munich: Oldenbourg.

Kater, Michael H. 2014. "Die Artamanen—Völkische Jugend in der Weimarer Republik." *Historische Zeitschrift* 213, no. 1: 577–638.

"Keine Staatsamateure mehr." 1951. *Der Spiegel*, December 25, 22.

Kellerhoff, Sven Felix. 2017. "'Ui, Herr Hitler, Sie dressieren Hunde aber fein!'" *Welt.de*, May 18. https://www.welt.de/geschichte/ plus164685882/Ui-Herr-Hitler -Sie-dressieren-Hunde-aber-fein.html.

Kelley, Douglas M. 1961. *22 Cells in Nuremberg: A Psychiatrist Examines the Nazi Criminals*. New York: MacFadden.

Kershaw, Ian. 1998. *Hitler, 1889–1936: Hubris*. London: Allen Lane.

Kershaw, Ian. 2001. *Hitler: 1936–1945: Nemesis*. London: Penguin.

Kircher-Kannemann, Anja. 2017. "Friedrich Schwangart—ein Mann zwischen Poesie und Wissenschaft." *Kultur-Geschichte(n)-Digital*, February 8. https://tour -de-kultur.de/2017/02/08/friedrich-schwangart-ein-mann-zwischen-poesie-und -wissenschaft.

Kirchmeier, Eva. 2013. "Annelies Umlauf-Lamatsch: Die Vielfalt in ihren Kinderbüchern." MA thesis, University of Vienna.

Klee, Ernst. 2009. *Das Kulturlexikon zum Dritten Reich: Wer war was vor und nach 1945*. Frankfurt am Main: Fischer Taschenbuch.

Klee, Ernst, Willi Dreßen, and Volker Rieß. 1988: *"Schöne Zeiten"—Judenmord aus Sicht der Täter und Gaffer*. Frankfurt am Main: Fischer.

Klemperer, Victor. 1947. *LTI: Notizbuch eines Philologen*. Berlin: Aufbau.

Klemperer, Victor. 1998. *I Shall Bear Witness*. Vol. 1 of *The Diaries of Victor Klemperer, 1933–41*. Translated by Martin Chalmers. London: Weidenfeld & Nicolson.

Klemperer, Victor. 1999a. *Ein Leben in Bildern*. Edited by Christian Borchert, Almut Giesecke, and Walter Nowojski. Berlin: Aufbau.

Klemperer, Victor. 1999b. *So sitze ich denn zwischen allen Stühlen: Tagebücher, 1945–1959*. 2 vols. Edited by Walter Nowojski and Christian Löser. Berlin: Aufbau.

Klemperer, Victor. 1999c. *To the Bitter End*. Vol. 2 of *The Diaries of Victor Klemperer, 1942–45*. Translated by Martin Chalmers. London: Weidenfeld & Nicolson.

Klemperer, Victor. 2006. *Language of the Third Reich: LTI—Lingua Tertii Imperii*. Translated by Martin Brady. London: Continuum.

Klemperer, Victor. 2015. *Ich will Zeugnis ablegen bis zum letzten*. 2 vols. Edited by Walter Nowojski and Hadwig Klemperer. Berlin: Aufbau.

Klemperer, Victor. 2017. *Warum soll man nicht auf bessere Zeiten hoffen? Ein Leben in Briefen*. Edited by Walter Nowojski and Nele Holdack. Berlin: Aufbau.

Klimenko, Iwan. 1965. "Wie ich die Leiche Hitlers fand." *Der Spiegel*, April 5.

Klingenberg, Axel. 2003. "Höllenkreis der Familie." *Jungle World*, June 25. https:// jungle.world/artikel/2003/26/hoellenkreis-der-familie.

Klothmann, Nastasja. 2015. *Gefühlswelten im Zoo: Eine Emotionsgeschichte, 1900– 1945*. Bielefeld: Transcript.

Klueting, Edeltraud. 2003. "Die gesetzlichen Regelungen der nationalsozialistischen Reichsregierung für den Tierschutz, den Naturschutz und den Umweltschutz." In *Naturschutz und Nationalsozialismus*, edited by Joachim Radkau and
Frank Uekötter, 77–105. Frankfurt am Main: Campus.

Knopf, Volker, and Stefan Martens. 2003. *Görings Reich: Selbstinszenierungen in
Carinhall.* Augsburg: Bechtermünz.

Kogon, Eugen. 1947. *Der SS-Staat: Das System der deutschen Konzentrationslager.*
Stockholm: Bermann-Fischer.

Kordt, Erich. 1950. *Nicht aus den Akten.* Stuttgart: Union Deutsche
Verlagsgesellschaft.

Koselleck, Reinhart. 2005. "Der Aufbruch in die Moderne oder das Ende des
Pferdezeitalters." In *Historikerpreis der Stadt Münster: Die Preisträger und Lau-
datoren von 1981 bis 2003*, edited by Berthold Tillmann, 159–74. Münster: Lit.

Köstering, Susanne, and Renate Rüb, eds. 2003. *Müll von gestern? Eine umweltge-
schichtliche Erkundung in Berlin und Brandenburg.* Münster: Waxmann.

Köstlin, Helmut. 1941. *Die Kartoffelkäferfibel.* Berlin: Deutsche Landwerbung/
Kartoffelkäfer-Abwehrdienst des Reichsnährstandes.

Krüger, Wilhelm. 1939. *Unser Pferd und seine Vorfahren.* Berlin: Springer.

Kube, Alfred. 1986. *Pour le mérite und Hakenkreuz: Hermann Göring im Dritten
Reich.* Munich: Oldenbourg.

Kube, Alfred. 1993. "Hermann Goering: Second Man in the Reich.'" In *The Nazi
Elite*, translated by Mary Fischer, edited by Ronald Smelser and Rainer Zitelmann, 62–73. New York: New York University Press.

Kuhnert, Max. 1993. *Will We See Tomorrow? A German Cavalryman at War, 1939–
1942.* London: Cooper.

Kutter, Kathrin Anna Maria. 2012. "Das Pferdebeschaffungswesen in der Bayerischen Armee von 1880–1920 an Hand der Akten des Kriegsarchives in
München." PhD diss., Ludwig-Maximilians-Universität, Munich.

Landeshauptstadt Dresden. 2010. *Abschlussbericht der Historikerkommission zu den
Luftangriffen auf Dresden zwischen dem 13. und 15. February 1945.* https://www
.dresden.de/media/pdf/infoblaetter/Historikerkommission_Dresden1945_Absch
lussbericht_V1_14a.pdf.

Laue, Christoph. 2015. "Jugend im Krieg." In *"Mit dem Führer zum Sieg?" Der
Raum Herford im Krieg, 1939–1945*, edited by Saskia Bruns und Christoph Laue,
40–50. Herford: Gedenkstätte Zellentrakt.

Leimann, Kaya. 2016. "Entstehung und Niedergang der Osnabrücker Seiden-
produktion—1920 bis 1945." Term paper, Universität Osnabrück. https://hvos
.hypotheses.org/files/2017/03/Entstehung-und-Niedergang-der-Osnabr%C3%BC
cker-Seidenproduktion-%E2%80%93-1920-bis-1945-1.pdf.

Leonhardt, Paul. 2008. "Die Wurzeln des Bundesjagdgesetzes." In *Jagdkultur—
gestern, heute, morgen: Symposium des Landesjagdverbandes Bayern e. V. und der
Bayerischen Akademie für Tierschutz, Umwelt- und Jagdwissenschaften, 18. und 19.
Juni 2008 in Rosenheim*, 35–44. Feldkirchen: Landesjagdverband Bayern.

Levi, Primo. 1985. "Small Causes." In *Moments of Reprieve: Essays*, 85–92. New York:
Simon & Schuster.

Levi, Primo. 1994. *Die dritte Seite: Essays und Erzählungen*. Munich: Deutscher Taschenbuch.

Lieb, Peter. 2007. *Konventioneller Krieg oder NS-Weltanschauungskrieg? Kriegführung und Partisanenbekämpfung in Frankreich 1943/44*. Munich: Oldenbourg.

Linse, Ulrich. 1993. "Der Film *Ewiger Wald*—oder, Die Überwindung der Zeit durch den Raum. Eine filmische Umsetzung von Rosenbergs 'Mythus des 20. Jahrhunderts.'" In *Formative Ästhetik im Nationalsozialismus: Intentionen, Medien und Praxis formen totalitärer ästhetischer Herrschaft und Beherrschung*, edited by Ulrich Herrmann and Ulrich Nassen, 57–75. Weinheim: Beltz.

Longerich, Peter. 2008. *Heinrich Himmler: Biographie*. Munich: Siedler.

Longerich, Peter. 2013. *Heinrich Himmler*. Translated by Jeremy Noakes and Lesley Sharpe. Oxford: Oxford University Press.

Longerich, Peter. 2016. *Wannseekonferenz: Der Weg zur "Endlösung."* Munich: Pantheon.

Lorenz, Konrad. 1942. *Die angeborenen Formen möglicher Erfahrung*. Berlin: Parey.

Lower, Wendy. 2013. *Hitler's Furies: German Women in the Nazi Killing Fields*. Boston: Houghton Mifflin Harcourt.

Lüdemann, Dagny. 2013. "Etiketten-Schwindel: Alles Wichtige zum Pferdefleisch -Betrug." *Die Zeit*, February 20. https://www.zeit.de/wissen/umwelt/2013-02/faq-pferdefleisch.

Lüttgenau, Rikola-Gunnar. 1993. "Ein Zoo in Buchenwald?" *Weimar-Kultur-Journal* 12:15–16.

Mackenzie, Compton. 1960. *Cat's Company*. London: Elek.

Maltzan, Maria von. 2009. *Schlage die Trommel und fürchte dich nicht: Erinnerungen*. Berlin: List.

Meibauer, Jörg. 2017. "'Um den Schädling zu vernichten': Propaganda, Hass, Humor und Metapher im Kindersachbuch: *Die Kartoffelkäferfibel* (1935) und *Karl Kahlfraß und sein Lieschen* (1952)." In *Verbale Aggression: Multidisziplinäre Zugänge zur verletzenden Macht der Sprache*, edited by Silvia Bonacchi, 289–303. Berlin: De Gruyter.

Meyer, Heinz. 1982. *Geschichte der Reiterkrieger*. Stuttgart: Kohlhammer.

Mildner, Horst, and Leonhard Resch. 1997. *Schorfheide zwischen Glanz und Entgleisung*. Schwedt: KIRO.

Misch, Rochus. 2014. *Hitler's Last Witness: The Memoirs of Hitler's Bodyguard*. London: Frontline.

Mohnhaupt, Jan. 2017. "Kamerad auf vier Tatzen." *JS-Magazin* 8:14–16. https://www.js-magazin.de/heftarchiv/2017.

Mohnhaupt, Jan. 2019. *The Zookeepers' War: An Incredible True Story from the Cold War*. Translated by Shelley Frisch. New York: Simon & Schuster.

Möhring, Maren. 2007. "'Hygienische Helfer': Katzen und Katzenschutz im nationalsozialistischen Deutschland." In *Von Katzen und Menschen: Sozialgeschichte auf leisen Sohlen*, edited by Clemens Wischermann, 173–82. Konstanz: UVK Verlagsgesellschaft.

Möhring, Maren. 2011. "'Herrentiere' und 'Untermenschen': Zu den Transformationen des Mensch-Tier-Verhältnisses im nationalsozialistischen Deutschland." *Historische Anthropologie* 19, no. 2: 229–44.

Motadel, David, Susanne Held, and Cathrine Hornung. 2017. *Für Prophet und Führer: Die islamische Welt und das Dritte Reich*. Stuttgart: Bundeszentrale für politische Bildung.

Müller, Inge. 1999. "Under the Rubble III." *Grand Street* 69:114.

Müller, Rolf-Dieter. 2005. *Der letzte deutsche Krieg: 1939–1945*. Stuttgart: Klett-Cotta.

Münkel, Daniela. 1996. *Nationalsozialistische Agrarpolitik und Bauernalltag*. Frankfurt am Main: Campus.

Münster, Erika. 2000. "Kurt Franz aus Ratingen." *Die Quecke: Ratinger und Angerländer Heimatblätter* 70 (December): 179–83.

Muth, Heinrich. 1970. "Hans Schlange-Schöningen (1886–1960)." In *Große Landwirte*, edited by Günther Franz and Heinz Haushofer, 394–417. Frankfurt am Main: DLG.

Neumärker, Uwe, and Volker Knopf. 2008. *Görings Revier: Jagd und Politik in der Rominter Heide*. Berlin: Links.

Nietzsche, Friedrich. 1974. *The Gay Science*. Translated by W. Kaufmann. New York: Vintage / Random House.

Oberthür, Wolfgang. 2002. *Schamesters Kriegsende*. Vol. 3 of *Der Schamester: Erinnerungen an eine Kindheit und Jugend auf dem Eichsfeld*. Duderstadt: Buchhandlung Mecke.

"Panjepferde." 1945. *Neue Zeit*, August 29, 3.

Passeick, Yannik. 2018. "Umweltschutz ist auch Heimatschutz? Was rechtsextreme Ideologien mit Natur- und Umweltschutz zu tun haben." *Rundbrief Forum Umwelt und Entwicklung* 4:30–31.

Perschke, Mario. 1999. "Der Hirsch von Johannes Darsow im Tierpark Berlin-Friedrichsfelde." *Mitteilungen des Vereins für die Geschichte Berlins* 95, no. 1: 481–85.

Perz, Bertrand. 1996. "'. . . müssen zu reißenden Bestien erzogen werden.' Der Einsatz von Hunden zur Bewachung in den Konzentrationslagern." *Dachauer Hefte: Studien und Dokumente zur Geschichte der nationalsozialistischen Konzentrationslager* 12:139–58.

Peterson, Agnes, and Bradley Smith, eds. 1974. *Heinrich Himmler: Geheimreden 1933 bis 1945 und andere Ansprachen*. Berlin: Propyläen.

Petzsch, Hans. 1964. "In Memoriam Friedrich Schwangart (1874–1958): Aus Anlaß der 90. Wiederkehr seines Geburtstages am 15. April 1964." *Anzeiger für Schädlingskunde* 37, no. 3: 60.

Peuschel, Harald. 1982. *Die Männer um Hitler. Braune Biographien: Martin Bormann, Joseph Goebbels, Hermann Göring, Reinhard Heydrich, Heinrich Himmler und andere*. Düsseldorf: Droste.

"Pferdelenker oder Pferdeschinder?" 1945. *Neue Zeit*, December 12, 3.

Picker, Henry. 1976. *Hitlers Tischgespräche im Führerhauptquartier*. Stuttgart: Seewald.

Piekalkiewicz, Janusz. 1992. *Pferd und Reiter im II. Weltkrieg*. Munich: Herbig.

Ploetz, Alfred. 1895. *Die Tüchtigkeit unsrer Rasse und der Schutz der Schwachen: Ein Versuch über Rassenhygiene und ihr Verhältniss zu den humanen Idealen, besonders zum Socialismus*. Berlin: Fischer.

Pollmer, Udo. 2015. "Der Schweinemord von 1915: Als die Wirtschaft eine Hungersnot provozierte." *Deutschlandfunk Kultur*, September 25. https://www.deutschlandfunkkultur.de/der-schweinemord-von-1915-als-die-wissenschaft-eine.993.de.html?dram:article_id=332117.

Pöppinghege, Rainer. 2009. "Abgesattelt!—Die publizistischen Rückzugsgefechte der deutschen Kavallerie seit 1918." In *Tiere im Krieg: Von der Antike bis zur Gegenwart*, edited by Rainer Pöppinghege, 234–50. Paderborn: Schöningh.

Pröse, Tim. 2016. *Jahrhundertzeugen: Die Botschaft der letzten Helden gegen Hitler. 18 Begegnungen*. Munich: Heyne.

"Quasi-Verrückte." 1994. *Der Spiegel*, November 13.

Räber, Hans. 2001. *Enzyklopädie der Rassehunde*. Vol. 1. Stuttgart: Kosmos.

Radkau, Joachim. 2002. *Natur und Macht: Eine Weltgeschichte der Umwelt*. Munich: Beck.

Rackow, Lutz. 2013. "Enttarnte Vorbilder." Interview with Zeitzeugen-Portal, Stiftung Haus der Geschichte der Bundesrepublik Deutschland, November 8.

Rahlf, Thomas, ed. 2015. *Deutschland in Daten: Zeitreihen zur Historischen Statistik*. Bonn: Bundeszentrale für politische Bildung.

Ramm, Eberhard, ed. 1922. *Deutsche Schweinehochzuchten*. Vol. 3 of *Deutsche Hochzuchten*. Berlin: Parey.

Rath, Martin. 2018. "'Dogmatik und tote Schweine.'" *Legal Tribune Online*, February 4. https://www.lto.de/recht/feuilleton/f/illegales-schlachten-strafrecht-nach-kriegszeit-gefaehrdung-bedarfsdeckung.

Raulff, Ulrich. 2015. *Das letzte Jahrhundert der Pferde: Geschichte einer Trennung*. Munich: Beck.

Raulff, Ulrich. 2018. *Farewell to the Horse: The Final Century of Our Relationship*. Translated by Ruth Ahmedzai Kemp. London: Penguin.

Reichholf, Josef H. 2014. *Auf den Hund gekommen*. Zürich: Vontobel Stiftung.

Reichsjugendführung, ed. 1938. *Pimpf im Dienst: Ein Handbuch für das Deutsche Jungvolk in der HJ*. Potsdam: Voggenreiter.

Reichsministerium des Innern. 1933. *Deutsches Reichsgesetzblatt*. 2 vols. Berlin: Reichsverlagsamt. http://alex.onb.ac.at/tab_dra.htm.

Reichsministerium des Innern. 1934. *Deutsches Reichsgesetzblatt*. 2 vols. Berlin: Reichsverlagsamt. http://alex.onb.ac.at/tab_dra.htm.

Reichsministerium des Innern. 1935. *Deutsches Reichsgesetzblatt*. 2 vols. Berlin: Reichsverlagsamt. http://alex.onb.ac.at/tab_dra.htm.

Reichsministerium desInnern. 1936. *Deutsches Reichsgesetzblatt*. 2 vols. Berlin: Reichsverlagsamt. http://alex.onb.ac.at/tab_dra.htm.

Reichsministerium des Innern. 1939. *Deutsches Reichsgesetzblatt*. 2 vols. Berlin: Reichsverlagsamt. http://alex.onb.ac.at/tab_dra.htm.

Reichsministerium des Innern. 1944. *Deutsches Reichsgesetzblatt*. 2 vols. Berlin: Reichsverlagsamt. http://alex.onb.ac.at/tab_dra.htm.

Reichsministerium für Volksaufklärung und Propaganda. 1941a. *Deutsche Wochenschau*, episode 561. https://archive.org/details/1941-06-04-Die-Deutsche-Wochenschau-561.

Reichsministerium für Volksaufklärung und Propaganda. 1941b. *Deutsche Wochen-schau*, episode 568. https://archive.org/details/1941-07-23-Die-Deutsche-Wochen schau-568.

Reichsministerium für Volksaufklärung und Propaganda. 1942. *Deutsche Wochen-schau*, episode 611. https://archive.org/details/1942-05-20-Die-Deutsche-Wochen schau-611.

Reichsministerium für Volksaufklärung und Propaganda. 1944. *Deutsche Wochen-schau*, episode 737. https://archive.org/details/1944-10-18-Die-Deutsche-Wochen schau-737.

Reichsministerium für Wissenschaft, Erziehung und Volksbildung. 1935. "För-derung des Seidenbaues (Sachsen)." *Deutsche Wissenschaft, Erziehung und Volksbildung: Amtsblatt des Reichsministeriums für Wissenschaft, Erziehung und Volksbildung und der Unterrichtsverwaltungen der Länder* 1, no. 21: 464.

Reichsministerium für Wissenschaft, Erziehung und Volksbildung. 1938. "För-derung des Seidenbaues (Sachsen)." *Deutsche Wissenschaft, Erziehung und Volksbildung: Amtsblatt des Reichsministeriums für Wissenschaft, Erziehung und Volksbildung und der Unterrichtsverwaltungen der Länder* 4, no. 10: 254–56.

Reichsministerium für Wissenschaft, Erziehung und Volksbildung. 1943. "För-derung des Seidenbaues." *Deutsche Wissenschaft, Erziehung und Volksbildung: Amtsblatt des Reichsministeriums für Wissenschaft, Erziehung und Volksbildung und der Unterrichtsverwaltungen der Länder* 9, no. 14: 225.

Reichsministerium für Wissenschaft, Erziehung und Volksbildung. 1945. "Seiden-bau." *Deutsche Wissenschaft, Erziehung und Volksbildung: Amtsblatt des Reichs-ministeriums für Wissenschaft, Erziehung und Volksbildung und der Unterrichts-verwaltungen der Länder* 11, no. 3: 16.

Reichsverband Deutscher Kleintierzüchter, ed. 1937. *Der Seidenbau in der Erzeu-gungsschlacht*. Berlin: Pfenningstorff.

Reischle, Hermann. 1935. *Reichsbauernführer Darré, der Kämpfer um Blut und Boden: Eine Lebensbeschreibung*. Berlin: Zeitgeschichte.

Rewald, Ilse. 1996. Interview by the Visual History Archive, USC Shoah Founda-tion, February 6. Transcribed by the Freie Universität Berlin. http://transcripts .vha.fu-berlin.de/interviews/49.

"Rezept: Wie hinterlasse ich sichtbare Spuren meiner national-sozialistischen Gesin-nung?" 1934. *HJ im Vormarsch*, August 4.

Riehl, Wilhelm Heinrich. 1854. *Die Naturgeschichte des Volkes als Grundlage einer deutschen SocialPolitik, Erster Band: Land und Leute*. Stuttgart: Cotta.

Roosevelt, Franklin D., and William C. Bullitt. 1972. *For the President, Personal and Secret: Correspondence between Franklin D. Roosevelt and William C. Bullitt*. Boston: Houghton Mifflin.

Roscher, Mieke. 2016. "Das nationalsozialistische Tier: Projektionen von Rasse und Reinheit im Dritten Reich." *Tierethik: Zeitschrift zur Tier-Mensch-Beziehung* 2, no. 13: 30–47.

Roscher, Mieke. 2018. "Praxeologische Einordnungen von Pferde- und Hundere krutierungen 1934–1944." *Hessische Blätter für Volks- und Kulturforschung* 53: 73–85.

Rubner, Heinrich. 1982. "Naturschutz, Forstwirtschaft und Umwelt in ihren Wechselbeziehungen, besonders im NS-Staat." In *Wirtschaftsentwicklung und Umweltbeeinflussung (14.–20. Jahrhundert)*, edited by Hermann Kellenbenz, 105–23. Wiesbaden: Steiner.

Rubner, Heinrich. 1997. *Deutsche Forstgeschichte, 1933–1945: Forstwirtschaft, Jagd und Umwelt im NS-Staat.* St. Katharinen: Scripta Mercaturae.

Rückerl, Adalbert, ed. 1977. *Nationalsozialistische Vernichtungslager im Spiegel deutscher Strafprozesse: Belzec, Sobibor, Treblinka, Chelmno.* Munich: Deutscher Taschenbuch.

Russell, Edward. 2008. *The Scourge of the Swastika: A Short History of Nazi War Crimes.* New York: Skyhorse.

Sächsische Landesanstalt für Landwirtschaft. 2003 *Zur Entwicklung der Schweinezucht und-produktion im Land Sachsen, 1850–2000.* Freistaat Sachsen: lebensministerium

Sambraus, Hans Hinrich. 1991. *Nutztierkunde: Biologie, Verhalten, Leistung und Tierschutz.* Stuttgart: Ulmer.

Sandner, Harald. 2016. *Hitler—Das Itinerar: Aufenthaltsorte und Reisen von 1889 bis 1945, Band I 1889–1927.* Berlin: Berlin Story.

Saraiva, Tiago. 2018. *Fascist Pigs: Technoscientific Organisms and the History of Fascism.* Cambridge, MA: MIT Press.

Sauerbruch, Ferdinand. 1979. *Das war mein Leben: Biographie.* Munich: Heyne.

Sax, Boria. 1997. "What Is a 'Jewish Dog'? Konrad Lorenz and the Cult of Wildness." *Society and Animals* 5, no. 1: 3–21.

Sax, Boria. 2017. *Animals in the Third Reich: Pets, Scapegoats and the Holocaust.* Pittsburgh: Yogh & Thorn.

Schaaf, Josef. 2011. "Weihnachten im Kessel von Stalingrad 1942." Interview with Zeitzeugen-Portal, Stiftung Haus der Geschichte der Bundesrepublik Deutschland, September 11. https://www.youtube.com/watch?v=wEh6bnqef2k.

Schäffer, Johann, and Lena König. 2015. "Der deutsche Tierschutz—ein Werk des Führers! Zum Umgang mit ideologisch kontaminierten Begriffen der NS-Zeit." *Deutsches Tierärzteblatt*, no. 9: 1244–56.

Schlange-Schöningen, Ernst-Siegfried. n.d. Unpublished book manuscript on the history of Schöningen family. Schlange-Schöningen family archives.

Schlange-Schöningen, Hans. 1946. *Am Tage danach.* Hamburg: Hammerich & Lesser.

Schlange-Schöningen, Hans. 1947. *Lebendige Landwirtschaft.* Hannover: Landbuch.

Schlange-Schoeningen, Hans. 1948. *The Morning After.* Translated by Edward Fitzgerald. London: Gollancz.

Schmitz-Berning, Cornelia. 2000. *Vokabular des Nationalsozialismus.* Berlin: De Gruyter.

Schmoller, Andreas. 2005. "Roberto Castellani: Erinnerungen anlässlich des ersten Todestages am 3. Dezember 2005." *Betrifft Widerstand*, no. 75: 30–33.

Schneider, Ernst. 1941. "'Die Erzeugungsschlacht wird fortgesetzt!'" *Zeitschrift für Schweinezucht*, no. 1: 1–2.

Schönfeld, Ana. 1996. Interview with the Visual History Archive, USC Shoah Foundation, January 17. Transcribed by the Freie Universität Berlin. http:// transcripts.vha.fu-berlin.de/interviews/32.

Schroeder, Christa, and Anton Joachimsthaler. 1985. *Er war mein Chef. Aus dem Nachlaß der Sekretärin von Adolf Hitler*. Munich: Langen Müller.

Schuller, Wolfgang. 2007. "Bastet, das Kätzchen: Die Katze im alten Ägypten." In *Von Katzen und Menschen: Sozialgeschichte auf leisen Sohlen*, edited by Clemens Wischermann, 13–24. Konstanz: UVK Verlagsgesellschaft.

Schwangart, Friedrich. 1937. *Vom Recht der Katze, mit Richtlinien für die Katzenhaltung*. Leipzig: Schöps.

Schwantje, Magnus. 1928. *Tierschlachtung und Krieg: Ein am 7. September 1927 auf dem VII. Demokratischen Friedens-Kongreß in Würzburg gehaltener Vortrag*. Berlin: Bund für radikale Ethik.

Schwantje, Magnus. 1942. *Sittliche Gründe gegen das Fleischessen*. Zürich: Mühlebach.

Seebo, Kurt W. 1998. "Der Schweinekrieg." In *Das Ende des Schweinekrieges: Marion Gölzows Photoaktion "Paarweise" als Friedenssymbol*, 5–13. Celle: Ströher.

Seidel, Günter. 2008. "Kindheit unter Hitler: 'Wir werden marschieren.'" *Spiegel Online*, January 20. http://www.spiegel.de/einestages/kindheit-unter-hitler-a-94 9037.html.

Selheim, Claudia. 2009. "'Alles wild'—Der edle Hirsch, ein gefragtes Bildmotiv." In *Vom Ansehen der Tiere*, 159–68. Nuremberg: Germanischen Nationalmuseum.

Semmler, Rudolf. 1947. *Goebbels, the Man Next to Hitler*. London: Westhouse.

Sontheimer, Michael. 2015. "Hitler in Bildchen." *Spiegel Online*, May 5. http://www .spiegel.de/einestages/adolf-hitler-bildbiografie-zeichnet-unreflektiertes-bild-a -1031935.html.

Speer, Albert. 1969. *Erinnerungen*. Berlin: Ullstein.

Speer, Albert. 2008. *Inside the Third Reich: Memoirs*. Translated by Richard and Clara Winston. New York: Simon & Schuster.

Spengler, Oswald. 1928. *Perspectives of World-History*. Vol. 2 of *The Decline of the West*. Translated by Charles Francis Atkinson. London: Allen & Unwin.

Spengler, Oswald. 1931. *Der Mensch und die Technik: Beitrag zu einer Philosophie des Lebens*. Munich: Beck.

Spengler, Oswald. 1932. *Man and Technics*. Translated by Charles Francis Atkinson. London: Allen & Unwin.

Spengler, Oswald. 1998. *Der Untergang des Abendlandes: Umrisse einer Morphologie der Weltgeschichte*. Munich: Beck.

Spiekermann, Uwe. 2018. *Künstliche Kost: Ernährung in Deutschland, 1840 bis heute*. Göttingen: Vandenhoeck & Ruprecht.

Springer, Hanns, and Rolf von Sonjevski-Jamrowski, dirs. 1936. *Ewiger Wald*. https://www.youtube.com/watch?v=1MJk3HUTtCU.

"Staatsstreich gegen die Wehrmacht." 1974. *Der Spiegel*, September 2, 92–102.

Stange, Thomas. 2015. "Den Tieren geht es gut." *Jungle World*, April 16. https:// jungle.world/artikel/2015/16/den-tieren-geht-es-gut.

Steinbach, Hans Joachim. 2010. "Der 'versteckte' Hirsch." *Wild und Hund* 3:100–101.

Stiftung Haus der Geschichte der Bundesrepublik Deutschland. 2011. *Die Hitler-Jugend und der BDM.* https://www.zeitzeugen-portal.de/videos/r5E7xhO1Z4w.

Strombeck, Friedrich Heinrich von, ed. 1829. *Ergänzungen des Allgemeinen Landrechts für die Preußischen Staaten.* Leipzig: Brockhaus.

Struck, Bernhard. 2015. "Schule im 'Dritten Reich.'" Deutsches Historisches Museum, Berlin, August 7. https://www.dhm.de/lemo/kapitel/ns-regime/alltagsleben/schule.html.

Stullich, Heiko. 2013. "Parasiten, eine Begriffsgeschichte." *Forum Interdisziplinäre Begriffsgeschichte* 2, no. 1: 21–9.

Świętorczecki, Bolesław. 1938. "Die Biologie des Wolfes." In *Waidwerk der Welt,* edited by the Reichsbund Deutsche Jägerschaft, 256–57. Berlin: Parey.

Syskowski, Hans M. F. 1996. *Rominter Heide.* Hamburg: Landsmannschaft Ostpreußen, Abteilung Kultur.

Taschwer, Paul. 2015. "Die verlorene Ehre des Konrad Lorenz." *Der Standard,* December 18. https://derstandard.at/2000027787429/Die-verlorene-Ehre- des -Konrad-Lorenz.

Tautorat, Hans-Georg. 1983. *Rominten.* Amended reprint. Hamburg: Landsmannschaft Ostpreußen, Abteilung Kultur.

Tautorat, Hans-Georg. 1996. "Ausflug in die Forst und Jagdgeschichte." In *Rominter Heide,* 10–12. Hamburg: Landsmannschaft Ostpreußen, Abteilung Kultur.

Thadeusz, Frank. 2011. "Tiere: Dick und doof." *Der Spiegel,* October 23. https://www.spiegel.de/spiegel/a-794054.html.

Thierfelder, Andreas. 1979. "Die antike Komödie und das Komische." *Würzburger Jahrbücher für die Altertumswissenschaft* 5:7–24.

Thissen, Torsten. 2014. "Der vergessene Prozess um Treblinka." *Rheinische Post,* October 19. https://rp-online.de/nrw/staedte/duesseldorf/der-vergessene-prozess -um-treblinka_aid-16537861.

Thünen-Institut. 2017. "Nutztierhaltung und Fleischproduktion in Deutschland." https://www.thuenen.de/de/thema/nutztiershyhaltung-und-aquakultur/ nutztierhaltung-und-fleischproduktion-in-deutschland.

Thünen-Institut. 2018. *Steckbriefe zur Tierhaltung in Deutschland: Mastschweine.* https://literatur.thuenen.de/digbib_extern/dn060514.pdf.

Tournier, Michel. 1972. *The Erl-King.* Translated from the French by Barbara Bray. New York: Doubleday.

Tournier, Michel. 1989. *Der Erlkönig.* Translated from the French by Hellmut Waller. Berlin: Aufbau.

Trakehner Verband. 2021. "Geschichte vor 1945." https://www.trakehner-verband .de/verband/geschichte.

Trittel, Günther J. 1987. "Hans Schlange-Schöningen: Ein vergessener Politiker der 'ersten Stunde.'" *Vierteljahreshefte für Zeitgeschichte* 35, no. 1: 25–63.

Tucholsky, Kurt. 2013. *Der Hund als Untergebener: Bissiges über Hunde und ihre Halter.* Großhansdorf: Officina Ludi.

Twardoch, Szczepan. 2017. *Drach*. Translated from the Polish by Olaf Kühn. Reinbek: Rowohlt Taschenbuch.

"Urmacher unerwünscht." 1954. *Der Spiegel*, June 22, 12–14.

Urmersbach, Viktoria. 2018. *Im Wald, da sind die Räuber: Eine Kulturgeschichte des Waldes*. Berlin: Vergangenheitsverlag.

Uther, Hans-Jörg. 2013. *Handbuch zu den "Kinder- und Hausmärchen" der Brüder Grimm: Entstehung—Wirkung—Interpretation*. Berlin: De Gruyter.

Van Vuure, Cis. 2005. *Retracing the Aurochs: History, Morphology and Ecology of an Extinct Wild Ox*. Sofia: Pensoft.

Varusschlacht im Osnabrücker Land. 2017. "Die Suche nach dem Ort." Museum und Park Kalkriese, Bramsche-Kalkriese. http://www.kalkriese-varusschlacht .de/varusschlacht/die-suche-nach-dem-ort.

Verein für Deutsche Schäferhunde. 1999. "100 Jahre: Der Deutsche Schäferhund." *SV-Zeitung*, special issue, April.

"Verhindert Transportverluste bei Schlachtschweinen!" 1940. *Zeitschrift für Schweinezucht*, no. 30: 238.

Vesper, Bernward. 1995. *Die Reise: Romanessay*. Reinbek: Rowohlt Taschenbuch.

"Vesper, Will." n.d. Munzinger Online/Personen—Internationales Biographisches Archiv. http://www.munzinger.de/document/00000004175.

Vinogradov, V. K., J. F. Pogonyi, and N. V. Teptzov, eds. 2005. *Hitlers Death: Russia's Last Great Secret from the Files of the KGB*. London: Chaucer.

von Preußen, Sibylle, and Friedrich Wilhelm von Preußen, eds. 2012. *Friedrich der Große: Vom anständigen Umgang mit Tieren*. Göttingen: MatrixMedia.

von Sagel, Irene, ed. 1981. *Justiz und NS-Verbrechen: Sammlung deutscher Strafurteile wegen nationalsozialistischer Tötungsverbrechen, 1945–1966*. Vol. 22. Amsterdam: Amsterdam University Press.

von Scheffel, Joseph Victor. 1864. "Lieder des Katers Hiddigeigei." In *Der Trompeter von Säckingen: Ein Sang vom Oberrhein*, 220–31. Zutphen: Thieme.

von Stephanitz, Max. 1921. *Der deutsche Schäferhund in Wort und Bild*. Jena: Kämpfe.

von Stephanitz, Max. 1923. *The German Shepherd Dog in Word and Image*. Translated by Carrington Charke. Jena: Kämpfe.

Vosganian, Varujan. 2017. *The Book of Whispers*. Translated by Alistair Ian Blyth. New Haven, CT: Yale University Press.

Walk, Joseph, ed. 2013. *Das Sonderrecht für die Juden im NS-Staat: Eine Sammlung der gesetzlichen Massnahmen und Richtlinien, Inhalt und Bedeutung*. Heidelberg: Müller 2013.

Weinberg, Gerhard L., ed. 2008. *Hitler's Table Talk, 1941–1944: His Private Conversations*. New updated ed. Translated by Norman Cameron and R. H. Stevens. New York: Enigma.

Weiser, Veronika. 2003. "15,000 Kokons für einen Fallschirm." *Kölner Stadt-Anzeiger*, January 30. https://www.ksta.de/15-000-kokons-fuer-einen-fallschirm -14593358.

Wiborg, Susanne. 2007. "Das Monster im Moor." *Die Zeit*, December 27.

Wippermann, Wolfgang. 1998. "Der Hund als Propaganda- und Terrorinstrument im Nationalsozialismus." In *Veterinärmedizin im Dritten Reich*, edited by Johann Schäffler, 193–206. Gießen: Deutschen Veterinärmedizinischen Gesellschaft.

Wippermann, Wolfgang, and Detlef Berentzen. 1999. *Die Deutschen und ihre Hunde: Ein Sonderweg der Mentalitätsgeschichte?* Munich: Siedler.

"Wo der Hund begraben liegt." 1948. *Der Spiegel*, March 9.

Wohlfromm, Hans-Jörg, and Gisela Wohlfromm. 2001. *Deckname Wolf: Hitlers letzter Sieg*. Berlin: Edition q.

Wohlfromm, Hans-Jörg, and Gisela Wohlfromm. 2017. *"Und morgen gibt es Hitlerwetter!" Alltägliches und Kurioses aus dem Dritten Reich*. Cologne: Anaconda.

Wowra, Walter. 1933. "Fettschwein oder Fleischschwein?" *Zeitschrift für Schweinezucht*, no. 13.

Wuketits, Franz M. 2011. *Schwein und Mensch: Die Geschichte einer Beziehung*. Hohenwarsleben: Westarp Wissenschaften.

Wunder, Michael. 2021. "Was heißt Eugenik?" Gedenkort-T4. https://www.gedenkort-t4.eu/de/wissen/was-heisst-eugenik.

Wylegalla, Reinhard. 2011. "Vom Maulbeerblatt zum Seidenkleid." *Deutsche Apotheker-Zeitung* 40:130. https://www.deutsche-apotheker-zeitung.de/daz-az/2011/daz-40-2011/vom-maulbeerblatt-zum-seidenkleid.

Zechner, Johannes. 2011. "Von 'deutschen Eichen' und 'ewigen Wäldern': Der Wald als nationalpolitische Projektionsfläche." In *Unter Bäumen: Die Deutschen und ihr Wald*, edited by Ursula Breymayer and Bernd Ulrich, 230–35. Berlin: Sandstein Kommunikation.

Zechner, Johannes. 2016. *Der deutsche Wald: Eine Ideengeschichte zwischen Poesie und Ideologie, 1800–1945*. Darmstadt: Philipp von Zabern.

Zechner, Johannes. 2017. "Natur der Nation. Der 'deutsche Wald' als Denkmuster und Weltanschauung." *Aus Politik und Zeitgeschichte*, nos. 49–50: 4–10.

Zelinger, Amir. 2018. *Menschen und Haustiere im Kaiserreich: Eine Beziehungsgeschichte*. Bielefeld: Transcript.

Zieger, Wilhelm. 1973. *Das deutsche Heeresveterinärwesen im Zweiten Weltkrieg*. Freiburg: Rombach.

Zipfel, Gaby. 2014. "'Wär' sie doch ein Stück von mir': Eva Klemperer in Victor Klemperers Tagebüchern." *Germanica*, no. 27: 41–58.

Zoller, Albert. 1949. *Hitler privat: Erlebnisbericht seiner Geheimsekretärin*. Düsseldorf: Droste.

Zorn, Wilhelm. 1963. *Schweinezucht: Züchtung, Fütterung, Haltung*. Edited by Gustav Comberg and Karl Richter. Stuttgart: Ulmer.

INDEX

Adorno, Theodor, 10–11
Airedale terriers: at concentration camps, 28; Nazi preference for, 20
Aktion Reinhardt (Operation Reinhard), 27
Ammon, Otto, 38
animal(s): de-extinction efforts, 99; human (*Menschentiere*), 6; humans reflecting qualities of, 3; interaction with, human relationships reflected in, 10–11; with "Jewish" characteristics, 9, 62, 63; master (*Herrentiere*), 6, 16; in Nazi ideology, 7–10, 21–22, 133–34; under Nazism, limited research on, 4, 8; *Nutztier* (working/utility), 9, 33, 34, 46; purebred, desirability of, 20; racial selective breeding of, 17–18, 19–20. *See also specific species*
animal breeding, and ideas about human races, 16, 19–20, 40
animal rights movement, early, 48
animal welfare, Nazi concern for, 49–50, 71–72; vs. crimes against humanity, 6–7, 50; legislation reflecting, 7, 49–50, 90–91; vs. treatment of cats, 73
antisemitism: in children's books, 62; and criticism of shehitah (ritual slaughter), 49–50; in film, 63; "scientification" of, Nazis and, 63

Arminius (Hermann the German), 100
Arndt, Ernst Moritz, 100
art: horse motif in, 114; red stag motif in, 86–87. *See also* statue(s)
Artaman League, 39
artists, *Gottbegnadeten-Liste* of, 125
Aryan race: domestication of pigs and, 40–41; German shepherd associated with, 133; identification of, 73–74; plan to renew, 40
aurochs: attempt to resurrect, 99; replicas of, at International Hunting Exhibition (1937), 22
Auschwitz concentration camp, 81; commandant of, 6–7, 28; laundry women at, resistance by, 132
Axmann, Artur, 64, 65–66

Backe, Herbert, 45, 46
Baltic countries, eugenics laws in, 19
Barry (dog), 26–27, 29–30
Bauer, Franz, 29
bear den, at Buchenwald Zoological Garden, 3, 4, 5; postcards depicting, 5; prisoners thrown into, 8
bee drones, Jews compared to, 62
Berghof compound, Hitler's, 24
Bergmann, Hans, 5, 6
Berlin: fall to Red Army, 128, 130; Goering's residence in, 79; Grunewald in, 92; International